ANDERSONIAN LIBRARY

WITHDRAWN
FROM
LIBRARY
STOCK

UNIVERSITY OF STRATHCLYDE

This book is to be returned on or before
the last date stamped below.

Catalysts for the Control of Automotive Pollutants

James E. McEvoy, *Chairman*

A symposium sponsored by the
Division of Industrial and Engineering
Chemistry and co-sponsored by the
Board-Council Committee on Chemistry and
Public Affairs, the Division of
Environmental Chemistry, the Division
of Fuel Chemistry, and the Division
of Petroleum Chemistry at the 167th
meeting of the American Chemical Society,
Los Angeles, Calif., April 2–4, 1974.

ADVANCES IN CHEMISTRY SERIES **143**

AMERICAN CHEMICAL SOCIETY
WASHINGTON, D. C.   1975

Library of Congress CIP Data

Catalysts for the control of automotive pollutants.
  (Advances in chemistry series; 143 ISSN 0065-2393)

Includes bibliographical references and index.

  1. Motor vehicles—Pollution control devices—Congresses. 2. Catalysts—Congresses.
  I. McEvoy, James E., 1920– II. American Chemical Society. Division of Industrial and Engineering Chemistry. III. American Chemical Society. IV. Series: Advances in chemistry series; 143.

QD1.A355 no. 143 [TL214.P6] 540'.8s 629.2'5
75-20298
ISBN 0-8412-0219-2   ADCSAJ 143 1–199 (1975)

Copyright © 1975

American Chemical Society

All Rights Reserved

PRINTED IN THE UNITED STATES OF AMERICA

Advances in Chemistry Series
Robert F. Gould, *Editor*

Advisory Board

Kenneth B. Bischoff

Edith M. Flanigen

Jesse C. H. Hwa

Phillip C. Kearney

Egon Matijević

Nina I. McClelland

Thomas J. Murphy

John B. Pfeiffer

Joseph V. Rodricks

FOREWORD

ADVANCES IN CHEMISTRY SERIES was founded in 1949 by the American Chemical Society as an outlet for symposia and collections of data in special areas of topical interest that could not be accommodated in the Society's journals. It provides a medium for symposia that would otherwise be fragmented, their papers distributed among several journals or not published at all. Papers are refereed critically according to ACS editorial standards and receive the careful attention and processing characteristic of ACS publications. Papers published in ADVANCES IN CHEMISTRY SERIES are original contributions not published elsewhere in whole or major part and include reports of research as well as reviews since symposia may embrace both types of presentation.

CONTENTS

Preface vii

1. Catalytic Reduction of Oxides of Nitrogen Emissions in Auto Exhaust Gas 1
   W. A. Mannion, K. Aykan, J. G. Cohn, C. E. Thompson, and J. J. Mooney

2. Thermocatalytic Detection of NO_x 14
   William B. Innes

3. The Relative Resistance of Noble Metal Catalysts to Thermal Deactivation 24
   G. R. Lester, J. F. Brennan, and James Hoekstra

4. Variation of Selectivity with Support Chemistry in NO_x Removal Catalysts 32
   Geoffrey E. Dolbear and Gwan Kim

5. Thermodynamic Interaction between Transition Metals and Simulated Auto Exhaust 39
   Howard D. Simpson

6. Platinum Catalysts for Exhaust Emission Control: The Mechanism of Catalyst Poisoning by Lead and Phosphorous Compounds 54
   G. J. K. Acres, B. J. Cooper, E. Shutt, and B. W. Malerbi

7. A Comparison of Platinum and Base Metal Oxidation Catalysts ... 72
   Gerald J. Barnes

8. Deposition and Distribution of Lead, Phosphorus, Calcium, Zinc, and Sulfur Poisons on Automobile Exhaust NO_x Catalysts 85
   Dennis P. McArthur

9. The Chemistry of Degradation in Automotive Emission Control Catalysts 103
   Richard L. Klimisch, Jerry C. Summers, and James C. Schlatter

10. The Optimum Distribution of Catalytic Material on Support Layers in Automotive Catalysis 116
    James Wei and E. Robert Becker

11. Oxidative Automotive Emission Control Catalysts—Selected Factors Affecting Catalyst Activity 133
    Louis C. Doelp, David W. Koester, and Maurice M. Mitchell, Jr.

12. Thermally Stable Carriers 147
    R. Gauguin, M. Graulier, and D. Papee

13. Spinel Solid Solution Catalysts for Automotive Applications 161
    C. A. Leech, III and L. E. Campbell
14. Oxidation of CO and C_2H_4 by Base Metal Catalysts Prepared on Honeycomb Supports 178
    J. T. Kummer

Index 193

PREFACE

The papers presented in this volume represent the intensive research and development effort that shifted into high gear in 1970 to develop systems for removing pollutants from auto exhaust, specifically carbon monoxide, hydrocarbons, and nitrogen oxides. In December of that year the Clean Air Act of 1970 was legislated which defined permissible concentration levels for these pollutants for succeeding years, and assigned the enforcement function to the Environmental Protection Agency. Compliance was the responsibility of the automobile manufacturers. The schedule for reduction in pollution levels was such that the auto companies were able to comply with the regulations without major engine or exhaust system modifications until the introduction of the 1975 Model Year vehicles. In consideration of the time constraints, the auto companies saw the catalytic converter as the most technically and economically viable system for meeting the required standards by the compliance deadline. The 1975 emission standards were eased somewhat in 1973; however, the so-called 1975 Interim Standards were sufficiently rigid to require exhaust system modifications which included the use of the catalytic converter for the majority of the vehicles planned for 1975 model year introduction.

This symposium was conceived at a meeting of the Industrial and Engineering Chemistry Program Committee at the 1972 Spring Meeting of the American Chemical Society in Boston. It seemed appropriate to docket the symposium for the Spring Meeting in 1974—a few months before the introduction of the 1975 vehicles—at Los Angeles, at that time known as the smog capital of the world. As Symposium Chairman I had many misgivings about the viability of the symposium later that week and during the succeeding week when the EPA-held hearings (initiated by a request for deferment of the standards by an auto company) in which all segments of industry involved in developing the catalytic converter were subpoenaed to testify. As a result of these hearings, considerable disagreement between the auto makers, catalyst companies, and environmentalists about the technical viability of the converter became apparent, and realistically it aroused concerns that catalytic converters of sufficient reliability to meet the standards could be developed by the enforcement date. Interest in a symposium on the catalytic removal of auto pollutants would be minimal if catalytic converters could not be developed in sufficient time to comply with the regulations. Either the standards would have to be relaxed or another solution to the emission problem would have to be developed.

The technical problems to be overcome were certainly formidable. The success of the symposium as well as the use of the converter on 1975 model year vehicles attest to the ingenuity and dedicated research effort put forth by the auto makers, catalyst companies, universities, and chemical and petroleum companies. The interaction of the representatives of these groups with the Environmental Protection Agency was also significant in that an appreciation of the EPA's assignment, as well as that of the auto companies, was brought into focus and resulted in a synergistic effort oriented to define a best possible solution to meet the 1975 deadline.

The papers included in this symposium cover the full gamut of problems that had to be addressed. The physical and chemical stability of the catalysts had to be significantly improved over known catalysts in order to meet the 50,000 mile life requirement prescribed by the regulations. The effects of catalyst poisons such as lead, sulfur, phosphorus, etc. were also critical in relation to the limits of deposition that could be tolerated while maintaining catalyst effectiveness. The nature of the catalyst support or substrate became significant in relation to its interaction with the metallic components of the catalyst—adherence, distribution, and reactivity at high temperature.

During the period of this intensive effort, I became personally acquainted with many of the authors, and also with many others who were highly involved in various aspects of the total program. I saw a dedication to achievement of a difficult goal which went far beyond the normal efforts which in my experience I have observed in research and development programs. It was disturbing to see the highly critical comments of the total program effort which would from time to time appear in press, authored by individuals or groups who had no understanding of the problems to be overcome. The fact that success was achieved, and that the converter is an integral part of 1975 model year vehicles, is a tribute to the creativity and long days contributed by the research and development community.

In conclusion, I would like specifically to acknowledge the assistance of J. M. Komarmy of the AC Division of the General Motors Corporation who worked closely with me in the organization of this symposium and provided the requisite liaison for symposium organization with the auto industry—so vitally needed for the success of the symposium. Without Mike's input this symposium would not have had the breadth and quality that was finally achieved.

JAMES E. McEVOY

Air Products and Chemicals, Inc.
Allentown, Pa.
February 1975

ns in Auto Exhaust Gas

Catalytic Reduction of Oxides of Nitrogen Emissions in Auto Exhaust Gas

W. A. MANNION, K. AYKAN, J. G. COHN, C. E. THOMPSON, and J. J. MOONEY

Engelhard Industries Division, Engelhard Minerals and Chemicals Corp., Menlo Park, Edison, N.J. 08817

This report discusses the progress made in developing catalysts for use in the control of nitrogen oxides from automotive exhaust. A stabilized supported ruthenium catalyst for use as the NO_x reduction bed in a dual-bed system was evaluated in laboratory bench tests and on an engine dynamometer with lead sterile and unleaded fuel. Catalyst exposure to high-temperature lean operation on an engine dynamometer revealed that the catalyst is not completely stable to such an environment. Initial evaluation of a three-way conversion (TWC) catalyst demonstrated that good conversion of carbon monoxide, hydrocarbons, and nitrogen oxides is maintained after exposure to elevated temperatures under oxidizing conditions using unleaded fuel.

Oxides of nitrogen are important components of photochemical smog which have an adverse effect on health (1). A major source of nitrogen oxides is the exhaust gas from internal combustion engines; this was recognized by the Clean Air Act of 1970 which requires effective control of such emissions. Some reduction in the output of nitrogen oxides by automobiles has been effected through a reduction in the peak temperature reached during combustion by diluting the cylinder charge with exhaust gases (exhaust gas recirculation) (2). This technique is of only limited effectiveness before driveability and fuel economy are affected too severely.

Catalysts

NO_x **Reduction Catalyst.** The use of catalysts to control emissions from automobiles has been investigated for a long time (3). However,

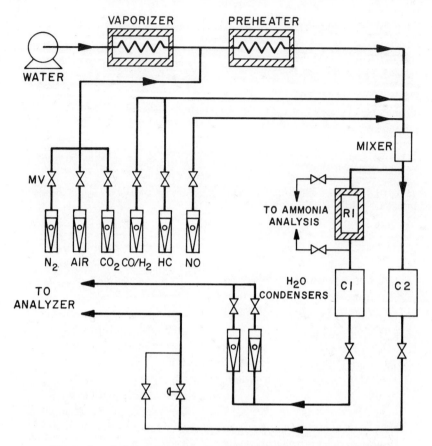

Figure 1. Schematic of bench scale NO_x catalyst screening unit

only lately has attention been focused on catalytic means of controlling oxides of nitrogen. In the dual-bed method, the internal combustion engine is operated net fuel rich, and a catalyst is used to promote the reduction of oxides of nitrogen to nitrogen with minimum ammonia formation. Air is then injected downstream of the reduction catalyst to make the stream net fuel lean, and the second bed catalyzes the oxidation of carbon monoxide and hydrocarbons (as well as any ammonia formed in the first bed). On warm-up, air is injected upstream of the reduction catalyst to control quickly the carbon monoxide and hydrocarbons by reaction in the manifold or over the reduction catalyst. The reduction catalyst, therefore, must be stable to repeated exposure to an oxidizing environment at high temperatures. A number of recent papers deal with the selection of a reduction catalyst (4, 5, 6, 7) and the durability problems that must be overcome (8, 9, 10, 11, 12). This paper reports on development of a supported ruthenium catalyst, stabilized by alloy forma-

Table I. Operating Conditions of NO_x Bench Scale Test Unit[a]

Gases	Amount, vol %	Analytical Method[b]
Constant		
NO	0.2	NDIR
HC ($3C_2H_4 + 2C_3H_8$)	0.03	FID
CO_2	13.5	—
H_2O	13.5	—
N_2	balance	—

	Amount (vol %) at Nominal A/F Ratio[c]					
	13.5	14.5	14.7	14.9	15.1	
Variable						
O_2	0.1	0.8	0.8	0.8	1.0	diffusion-cond. cell
CO	3.0	1.5	1.0	0.7	0.5	NDIR
H_2	1.0	0.5	0.33	0.23	0.17	—

[a] Temperature: variable to 650°C; VHSV: variable to 200,000/hr.
[b] NDIR: non-dispersive IR; FID: flame ionization detector.
[c] Stoichiometric nominal A/F is 14.65. Ignition tests were run at a constant nominal A/F ratio.

tion against ruthenium loss under oxidizing conditions, for use as the reduction catalyst in a dual-bed system.

Three-Way Conversion (TWC) Catalyst. The dual-bed method requires net fuel-rich operation of the engine and therefore lower fuel economy. Interest has heightened recently in a technique to control all three pollutants—carbon monoxide, hydrocarbon, and nitrogen oxides—by maintaining the air/fuel (A/F) at near-stoichiometric ratios and by converting all pollutants over a single catalyst. A sensor is used to control the partial pressure of oxygen in the exhaust stream by feedback to a carburetion or fuel injection system (*13*). The increased cost of the sensor–carburetor system will be compensated for by better fuel economy,

Table II. Engine Aging of Catalysts[a]

Test	Fuel	Air/Fuel Ratio	Time, hrs	Catalysts Tested	Designation
1	Lead sterile	13.9	48	no	—
	<0.003 g Pb/gal,	15.1	6	no	—
	25 ppm S,	13.9	6	yes	1–60
	<1 ppm P	16.1	8.5	yes	1–68
2	Durability	13.9	48	yes	2–48
	0.035 g Pb/gal,	15.1	6	no	—
	340 ppm S,	13.9	6	yes	2–60
	<1 ppm P	16.4	8	yes	2–68

[a] New catalysts were used for each test. Inlet temperature, 760°C; engine dynamometer operated at steady state, 57 mph under a road load.

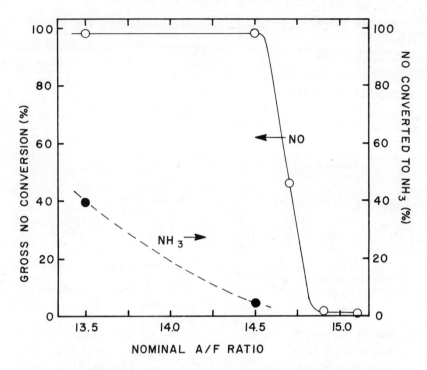

Figure 2. Performance of virgin, supported ruthenium catalyst for removal of nitric oxide from a synthetic exhaust

Test conditions on laboratory unit: temperature, 500°C; and VHSV, 200,000/hr

the absence of an air pump, and the use of a single catalyst. Catalysts have been developed for this application, and a preliminary evaluation is given.

Experimental

Catalysts for both the applications described above have been prepared on particulate and monolithic supports. For this investigation, all catalysts were prepared on a commercially available, monolithic support from American Lava. A coating of high surface area, stabilized alumina was applied to the ceramic, and it was then impregnated. Ruthenium or a ruthenium alloy catalyst (Engelhard N-2) was used for the reduction catalyst whereas ruthenium was not used for the TWC catalyst (TWC-1). In all preparations containing ruthenium, the loading was such that annual availability was not exceeded based on the assumption that all cars sold in the United States contained the catalyst.

A test unit was constructed for evaluation of the catalysts (Figure 1). The feed gas contained the significant components of auto exhaust (*see* Table I). Only nitric oxide, NO, was used because it is the oxide of nitrogen most prevalent in auto exhaust gas. The catalysts were tested at each selected catalyst inlet temperature under conditions ranging from reducing to oxidizing by holding the concentration of most of the gases constant while varying the amounts of carbon monoxide, hydrogen, and oxygen. The nominal air/fuel ratios were calculated by the Eltinge method (14). After stabilization at a selected condition, conversions were determined by measuring upstream and downstream concentrations. Catalysts could be tested on this unit at volume hourly space velocities (VHSV) of up to 2×10^5/hr. All preparations were made on cylindrical, monolithic supports 1.5 in. in diameter and 3 in. long. In order to test these pieces at a space velocity of 2×10^5/hr, a ceramic ring with a hole 0.75 in. in diameter was placed over the pieces. Ammonia was analyzed before and after the catalyst by absorption in dilute sulfuric acid and titration.

Catalysts were aged either in the laboratory or on an engine to aid in the screening process. In the laboratory aging, a test cylinder 1.5 in. in diameter and 1.5 in. long was placed in a tube furnace and heated to 865°C under flowing nitrogen. Then the gases listed in Table I, except hydrogen, were passed over the catalyst at a net oxidizing condition of 14.7 nominal air/fuel ratio and a space velocity of 19,200/hr. Tempera-

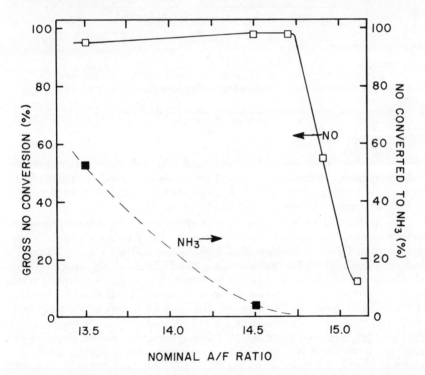

Figure 3. Performance of virgin, stabilized support ruthenium catalyst (N-2) for removal of nitric oxide from a synthetic exhaust

Test conditions on laboratory unit: temperature, 500°C; and VHSV, 200,000/hr

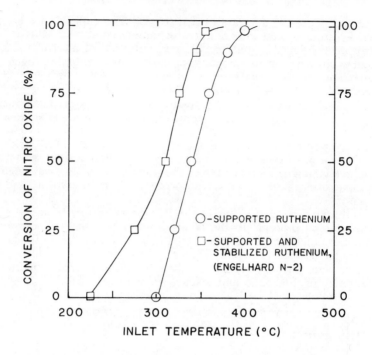

Figure 4. Ignition performance of virgin, supported ruthenium and stabilized, supported ruthenium (N-2) catalysts

Test conditions on laboratory unit: nominal A/F ratio, 14.5/1; and VHSV, 200,000/hr

ture control was maintained by a sheathed chromel–alumel thermocouple just downstream of the catalyst. Test pieces were heated at this condition for various lengths of time, cooled under nitrogen, and then analyzed for ruthenium content by x-ray fluorescence.

An additional controlled aging was effected on an engine dynamometer operating at steady state with lead sterile (< 0.003 g Pb/gal, 25 ppm S, < 1 ppm P) or durability (0.035 g Pb/gal, 340 ppm S, < 1 ppm P) fuel (*see* Table II). The engine was a Ford 351C cid V8 equipped with needle valves on the main jets of the carburetor so that conditions could be made reducing or oxidizing. The A/F ratio was checked by analysis of the exhaust gases (*14*). The exhaust manifolds and pipes were insulated, and the engine dynamometer was operated at 57 mph under a road load so that inlet temperature to the reactors was about 760°C. The catalysts were aged in multichamber reactors. Each reactor contained four test pieces, 1.5 in. in diameter and 3 in. long, that could be removed and tested in the laboratory test unit and then returned to the engine for further aging. Uncatalyzed monolith was also placed in each reactor so that the total area open to flow was approximately 17 in.² per reactor. This ensured that the test pieces were exposed to a volume of exhaust gas during the test that was similar to that expected in actual use, assuming a catalyst volume of 51 in.³ per bank of four cylinders.

Results and Discussion

NO$_x$ Reduction Catalyst. One of the most promising catalysts for reducing nitric oxide is ruthenium (Figure 2). It is unusual in that it catalyzes the reduction of NO with a low selectivity to ammonia which, if present, can be reoxidized in the second bed of a dual-bed system. When it is exposed at high temperatures to an oxidizing environment, however, ruthenium is lost through the formation of the volatile tetroxide.

As a result of efforts to stabilize the ruthenium, a new formulation was developed, the N-2 catalyst. This catalyst has good virgin activity for removing nitric oxide, even in a slightly oxidizing environment (Figure 3). More ammonia was formed by this catalyst at 500°C than by the ruthenium catalyst at very rich A/F ratios. Aging in a reducing environment substantially reduced the selectivity to ammonia (*see* figures below).

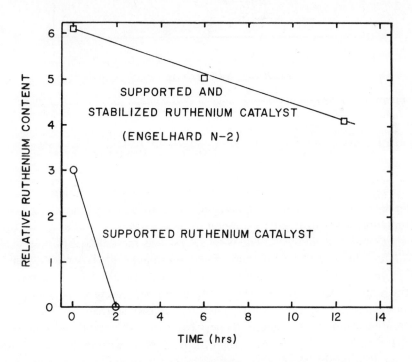

Figure 5. Ruthenium loss under oxidizing conditions with a synthetic blend

Conditions: temperature, 865°C; nominal A/F, 14.7 (slightly oxidizing); and VHSV, 19,200/hr

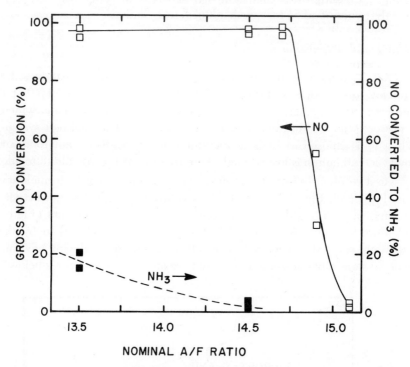

Figure 6. Performance of engine-aged N-2 catalyst for removal of nitric oxide from a synthetic exhaust.

Aging conditions on dynamometer: fuel, lead sterile (<0.003 g Pb/gal, 25 ppm S, <1 ppm P); and time, 60 hrs 10% oxidizing (A/F, 15/1). Test conditions on laboratory unit: temperature, 500°C; and VHSV, 100,000/hr.

Since the warm-up characteristic of a catalyst is an important consideration in auto exhaust application, the conversion of nitric oxide as a function of catalyst inlet temperature was studied for the virgin ruthenium and N-2 catalysts. The N-2 catalyst was more active (Figure 4).

The rate of loss of ruthenium was determined after laboratory aging in a synthetic, slightly oxidizing environment (Figure 5). The apparently higher ruthenium content of the stabilized catalyst was partially compensated for by a difference in ceramic density so that the ruthenium content per unit volume was similar. Note that the stabilized catalyst, N-2, lost ruthenium more slowly than the unstabilized catalyst. The linear rate of loss of the stabilized catalyst indicates that the ruthenium loss was zero order in ruthenium. For these specific conditions, a significant improvement in stabilizing the ruthenium to an oxidizing atmosphere was achieved by alloy formation.

The N-2 catalyst was aged on the engine dynamometer. Two agings were performed, each with a new pair of N-2 catalysts. The aging conditions are listed in Table II. The initial aging was with lead sterile fuel, and the catalysts were first removed after 60 hrs (3400 miles), 10% of which was oxidizing (test 1-60). In Figure 6 are plotted the data from the pair of catalysts tested at 500°C inlet temperature and 100,000/hr space velocity. The catalysts still appeared quite active so they were returned to the engine dynamometer for an additional 8.5 hrs under a stronger oxidizing A/F ratio (test 1-68). Figure 7 demonstrates the effect of this aging on NO conversion and selectivity to ammonia. Deterioration in the net conversion was apparent.

A second pair of N-2 catalysts were aged on the engine dynamometer using durability fuel. After 48 hrs under reducing conditions (test 2-48), there was little effect of lead for the first 2700 miles (Figure 8). After an additional 6 hrs under oxidizing conditions and 6 hrs under reducing conditions (test 2-60) deterioration was evident (Figure 9). One of the

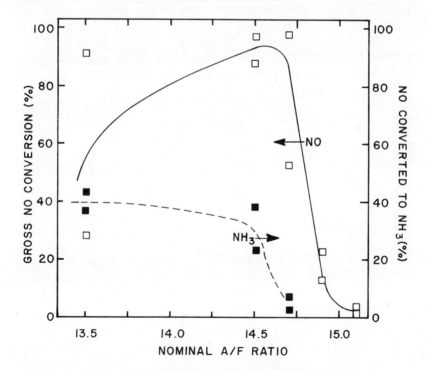

Figure 7. Performance of engine-aged N-2 catalyst for removal of nitric oxide from a synthetic exhaust.

Aging conditions on dynamometer: fuel, lead sterile; and time, 60 hrs 10% oxidizing (A/F, 15/1) + 8.5 hrs oxidizing (A/F, 16/1). Test conditions on laboratory unit: temperature, 500°C; and VHSV, 100,000/hr.

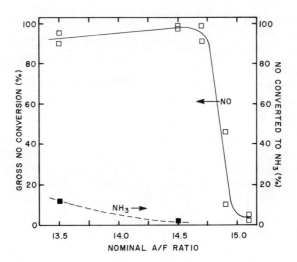

Figure 8. Performance of engine-aged N-2 catalyst for removal of nitric oxide from a synthetic exhaust.

Aging conditions on dynamometer: fuel, durability (0.035 g Pb/gal, 340 ppm S); and time, 48 hrs all reducing (A/F, 13.9/1). Test conditions on laboratory unit: temperature, 500°C; and VHSV, 100,000/hr.

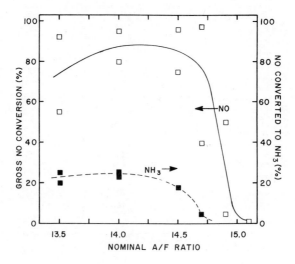

Figure 9. Performance of engine-aged N2- catalyst for removal of nitric oxide from a synthetic exhaust.

Aging conditions on dynamometer: fuel, durability; and time, 60 hrs 10% oxidizing (A/F, 15/1). Test conditions on laboratory unit: temperature, 500°C; and VHSV, 100,000/hr.

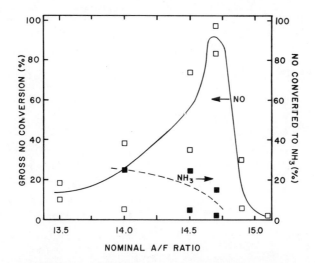

Figure 10. Performance of engine-aged N-2 catalyst for removal of nitric oxide from a synthetic exhaust.

Aging conditions on dynamometer: fuel, durability; and time, 60 hrs 10% oxidizing (A/F, 15/1) + 8 hrs oxidizing (A/F, 16.4/1). Test conditions on laboratory unit: temperature, 500°C; and VHSV, 100,000/hr.

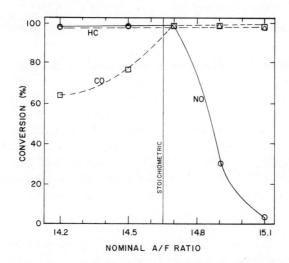

Figure 11. Performance of virgin TWC-1 catalyst for removal of carbon monoxide, hydrocarbons, and nitric oxide from a synthetic exhaust

Test conditions on laboratory unit: temperature, 650°C; and VHSV, 100,000/hr

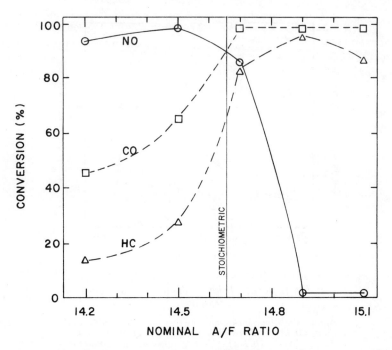

Figure 12. Performance of engine-aged TWC-1 catalyst for removal of three pollutants from a synthetic exhaust.

Aging conditions on dynamometer: fuel, durability; and time, 60 hrs 10% oxidizing (A/F, 15/1) + 8 hrs oxidizing (A/F, 16.4/1). Test conditions on laboratory unit: temperature, 650°C; and VHSV, 100,000/hr.

pair of catalysts lost more activity than the other, possibly as the result of differences in gas composition between the two cylinder banks of the V8 engine. Between A/F ratios of 14.0 and 14.5, conversion of NO was approximately 80% which was still acceptable. The catalysts were returned for a final 8 hrs under strongly oxidizing conditions (test 2-68) and then tested (see Figure 10). However, the ruthenium alloy was not sufficiently stable to withstand such an environment for an extended period of time. The mode of deactivation, where ammonia selectivity increased and NO conversion at rich A/F ratios decreased, was indicative of loss of ruthenium.

Three-Way Conversion (TWC) Catalyst. TWC-1 was evaluated on the same test unit with the catalyst inlet temperature raised to 650°C. The conversion of all three pollutants over a virgin catalyst is shown in Figure 11. The conversions of carbon monoxide and hydrocarbon at nominal A/F ratios much less than stoichiometric must be attributable mostly to steam reforming since the oxygen available was not sufficient to account for the observed conversion levels. This activity was lost after only a short time, but this is not important when the A/F ratio is stoichiometric.

TWC-1 catalyst was aged on the engine dynamometer with the N-2 preparations using durability fuel (test 2 in Table II). Evaluation of this catalyst after 68 hrs (test 2-68) is given in Figure 12. Good conversion of the three pollutants was maintained after this severe aging; the catalyst retained much of its activity. A means of aging the TWC-1 catalyst under conditions more closely approximating those of actual use is being developed.

Acknowledgment

All analyses were performed by the Instrumental Analysis Laboratory, Engelhard Industries Division, Engelhard Minerals and Chemicals Corp., Newark, N. J.

Literature Cited

1. *Fed. Regist.* (June 8, 1973) **38** (110), 151, 80.
2. Chipman, J. C., Chao, J. Y., Ingels, R. M., Jewell, R. G., Deeter, W. F., Soc. Auto. Eng., Nat. Auto. Eng. Mtg., Detroit, May, 1972, paper **720511**.
3. McDermott, J., "Catalytic Conversion of Automobile Exhaust 1971," p. 208, Noyes Data Corp., Park Ridge, N.J.
4. Klimisch, R. L., Taylor, K. C., *Environmental Sci. Technol.* (1973) **7**, 127.
5. Shelef, M., Gandhi, H. S., *Ind. Eng. Chem. Prod. Res. Develop.* (1972) **11**, 2.
6. *Ibid.* (1972) **11**, 392.
7. Lunt, R. S., Bernstein, L. S., Hansel, J. G., Holt, E. O., Soc. Auto. Eng., Auto. Eng. Cong., Detroit, January, 1972, paper **720209**.
8. Roth, T. F., *Ind. Eng. Chem. Prod. Res. Develop.* (1971) **10**, 381.
9. Meguerian, G. H., Hirschberg, E. H., Rakowsky, F. W., Lang, C. R., Shock, D. N., Soc. Auto. Eng., Nat. Auto. Eng. Mtg., Detroit, May, 1972, paper **720480** (SP. 370).
10. Aykan, K., Mannion, W. A., Mooney, J. J., Hoyer, R. D., Soc. Auto. Eng., Nat. Auto. Eng. Mtg., Detroit, May, 1973, paper **730592**.
11. Jackson, H. R., McArthur, D. P., Simpson, H. D., Soc. Auto. Eng., Nat. Auto. Eng. Mtg., Detroit, May, 1973, paper **730568**.
12. Neal, A. H., Wigg, E. E., Holt, E. L., Soc. Auto. Eng., Nat. Auto. Eng. Mtg., Detroit, May, 1973, paper **730593**.
13. Fleming, W. J., Howarth, D. S., Eddy, D. S., Soc. Auto. Eng., Nat. Auto. Eng. Mtg., Detroit, May, 1973, paper **730575**.
14 Eltinge, L., "Fuel-Air Ratio and Distribution from Exhaust Gas Composition," Soc. Auto. Eng., Auto. Eng. Cong., Detroit, January, 1968.

RECEIVED June 13, 1974.

2

Thermocatalytic Detection of NO_x

WILLIAM B. INNES

Purad Inc., Upland, Calif. 91786

Thermocatalytic detection by catalyst bed temperature rise was applied to NO_x using V_2O_5–Al_2O_3 catalyst. This involved the highly exothermic reaction of NH_3 with NO and O_2 to produce H_2O and N_2 or N_2O. Detector response was proportional to NO level when $NH_3/NO > 2$. Flow rates of 1–2 ft^3/hr with .03 cm^3 catalyst at 300–400°C seemed to be optimum. Higher oxides of nitrogen than NO, and their hydrates, responded like NO whereas N_2O and SO_2 gave negligible responses. The method has application in measuring NO_x directly in most sources. HC, CO, and NO_x in vehicle exhaust can be analyzed by series flow through V_2O_5–Al_2O_3 catalyst (HC), V_2O_5–Al_2O_3 bed (further HC removal), and Pt catalyst (CO detection), followed by NH_3 introduction and V_2O_5–Al_2O_3 catalyst (NO_x detection).

Thermocatalytic detection involving measurement of temperature rise in catalytic beds has been successfully used for analysis of HC and CO in exhaust gas (1). The objective of this study was to extend the approach to NO_x; this entailed finding a suitable exothermic, selective, catalytic reaction involving NO in the presence of oxygen as well as the other constituents of exhaust gas.

A review of heat effects at standard state conditions (25°C, 1 atm) to determine suitable reactions of NO yielded the following:

$$NO + CO \rightarrow CO_2 + 1/2\ N_2 + 89\ \text{kcal} \qquad (1)$$

$$NO + 1/2\ O_2 \rightarrow NO_2 + 14\ \text{kcal} \qquad (2)$$

$$NO + H_2 \rightarrow 1/2\ N_2 + H_2O + 80\ \text{kcal} \qquad (3)$$

$$NO + O_3 \rightarrow NO_2 + O_2 + 48\ \text{kcal} \qquad (4)$$

$$NO + 2/3\ NH_3 \rightarrow 5/6\ N_2 + H_2O + 72\ \text{kcal} \tag{5}$$

$$NO + 2\ NH_3 + O_2 \rightarrow 1\text{-}1/2\ N_2 + 3\ H_2O + 173\ \text{kcal} \tag{6}$$

$$NO + 2\ NH_3 + 1\text{-}3/4\ O_2 \rightarrow 1\text{-}1/2\ N_2O + 3\ H_2O + 144\ \text{kcal} \tag{7}$$

The use of ozone (Reaction 4) would require an ozone generator, and heat effects from ozone decomposition would probably interfere. The oxidation of NO to NO_2 with oxygen (Reaction 2) is reversible and has a low heat effect. Reactions with NO involving CO or H_2 (Reactions 1, 3) would necessitate introducing CO or H_2 into the gas stream to ensure its presence; furthermore, the catalyst could not effect oxidation of this gas in the presence of oxygen. There have been fairly extensive studies of numerous catalysts for these reactions; Shelef (2), for example, demonstrated that in the presence of excess oxygen the oxidation of CO or H_2 is generally very competitive and would seriously interfere.

Interaction of nitric oxide with ammonia can be highly exothermic. When Reactions 6 and 7 are compared on a C vs. N basis, the heat effect is of the same magnitude as hydrocarbon oxidation (173 kcal per NO by Reaction 6 vs. 160 kcal for typical hydrocarbon oxidation).

Since exploratory experimental work on the interaction of NO with NH_3 in air using vanadia–alumina catalyst indicated a large catalyst bed temperature rise, effort was concentrated on this system. Other oxidation catalysts appeared less attractive than vanadia because their known activity for CO and H_2 oxidation could cause interference from these

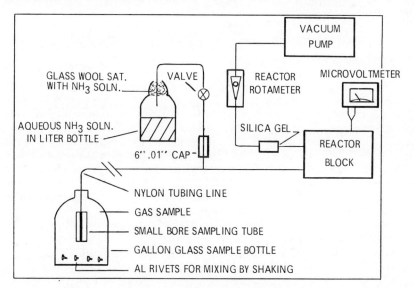

Figure 1. Schematic of system

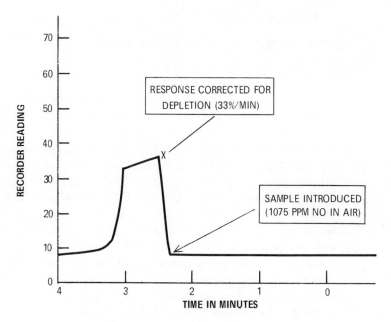

Figure 2. Representative NO response chart
Flow rate, 0.9 ft³/hr; and T, 395°

constituents unless a prior removal step was used. Furthermore, most active catalysts for these reactions, *e.g.* those containing platinum and copper, are active for ammonia oxidation.

Experimental Methods

The system used in these studies is depicted in Figure 1. Sample gas was syringe-injected into a gallon bottle containing room air. The sampling probe was then inserted in this bottle; ammonia could also be added *via* a capillary to the gas stream. After preheating, the mixture passed through the catalytic reactor containing vanadia catalyst and a glass-coated chromel–constantan thermocouple detector into the vacuum pump. Detector output was measured with a Keithley model 150B microvoltmeter. A reading of 80 μV corresponded to about a 1°C rise in temperature. Because of sample depletion with time, corrections were applied to the data (*see* typical response curve in Figure 2). In general, the data were as reproducible as the accuracy of preparing test samples and measuring the signal (about 2%).

Studies were made over a range of temperatures, flow rates, and NH_3/NO ratios. The position of the thermocouple probe tip in the catalyst bed was fixed at 0.125 in. of catalyst ahead of the tip. Catalyst bed diameter was 0.128 in., the volume 0.026 cm³. The vanadia–alumina catalyst was the same used for prior work on HC oxidation (*1*). Ammonia was introduced by premixing with NO in the sample bottle and by continuous addition *via* the capillary. Both aqueous ammonia vapor and cylinder NH_3 served as ammonia sources during continuous addition, whereas pure NH_3 was used for the premixtures. There was no evidence that the presence of some water vapor from the aqueous ammonia source affected the findings.

Results

Data for the premixtures at 395°C and 1.75 ft³/hr demonstrate the effect of ammonia level (Figure 3). There was some heat effect from ammonia alone at 395°C. Response to NO levels out at $NH_3/NO = 2$ after a correction is made for the ammonia response. Figure 4 indicates that the response corrected for ammonia response was linear with the NO level up to $NO/NH_3 = 0.5$. Nitric oxide was not detected in the product gas from the ammonia–catalyst–air interaction so it can probably be assumed that the nitrogen product was N_2O or N_2. This was not surprising because of the high rate for Reactions 6 or 7 *vs.* ammonia oxidation. That is, if NO formed, it would react with the excess NH_3.

A complex response curve was obtained when NH_3–NO mixes were tested at 220°C with full equilibrium requiring about a minute. This time effect and other evidence suggested that chemisorption of ammonia,

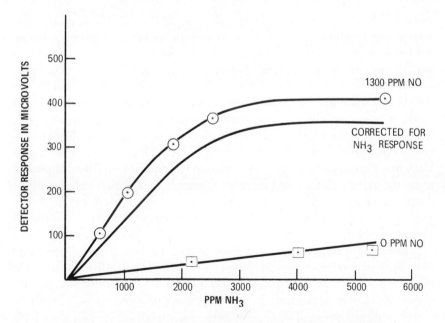

Figure 3. Effect of NH_3 level on response at fixed NO level
Both components in sample; flow rate, 1.7 ft³/hr; and T, 395°C

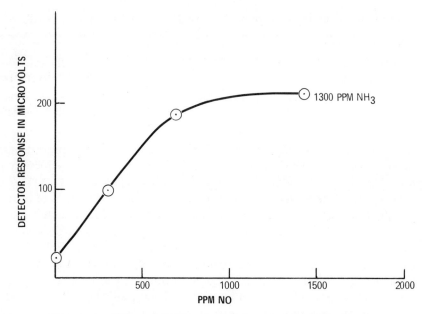

Figure 4. Effect of NO level on response at fixed NH_3 level
Both components in sample; flow rate, 1.7 ft³/hr; and T, 395°C

presumably by the alumina base (slow at 220°C), is a necessary first step in the reaction. There was no noticeable heat effect from ammonia oxidation at 220°C, but substantial transient heat effects from adsorption and desorption were apparent (about 15% of that for NO–NH_3 interaction).

When ammonia was added continuously, which would be the normal mode of operation, the response curve was generally like that in Figure 2, even at low temperatures. Figure 5 demonstrates that equilibrium response was linear with the NO level up to NO/NH_3 = 0.5 at 220°C and 1 ft³/hr. Equilibrium values for continuous ammonia addition appeared to be the same as those for premixed ammonia and nitric oxide provided a correction was applied for ammonia response at elevated temperatures. With continuous ammonia addition, this ammonia effect was nullified electronically.

Effect of Temperature and Flow Rate. Figures 6 and 7 illustrate the effect of temperature and flow rate on response. All findings were consistent with a highly exothermic, first order reaction with a high rate above 300°C. It is interesting that response appeared to be insensitive to temperature above 275°C at a flow rate of 1.1 ft³/hr and that high flow rates can be used to realize maximum response with minimum dependence on this parameter.

Magnitude of Response vs. That for HC Oxidation and Stoichiometry. The response of the NO–NH_3 interaction in the presence of excess oxygen and that of the highly reactive 1-butene oxidation were compared at 390°C at flow rates of 1.0 and 2.3 ft^3/hr. On a C vs. N basis, the C/N relative response was 1.27 at 1.0 ft^3/hr and 0.90 at 2.2 ft^3/hr. Response for HC's less reactive than butenes (e.g. butanes) was much less. The interaction of NO and NH_3 in the presence of oxygen apparently releases about the same heat and is more rapid than 1-butene oxidation. Reactions 6 and 7 both qualify in heat effect and ratio of NH_3/NO utilized. It seems probable that both reactions are involved. For our purposes, it was not necessary to make the distinction.

Some work was done to define the reaction better in which oxygen-free nitrogen was substituted for air and controlled amounts of O_2, NO, and NH_3 were added to the sample bottle. This work demonstrated that the oxygen level could be reduced to O_2/NO = 1.5 by using NO levels of 0.13 and 0.42% and 2/1 NH_3/NO at 390°C and 1 ft^3/hr with small affect on response for several minutes. However, addition of air to the sample during testing gave noticeable positive response, as would be expected if Reaction 6 or 7 were involved. Complete removal of oxygen would be expected to change the nature (valence state) of the catalyst

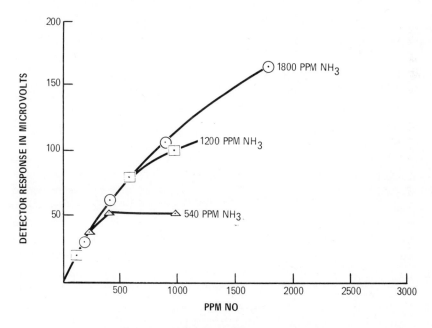

Figure 5. Effects of NO and NH_3 levels

Continuous NH_3 addition; flow rate, 1 ft^3/hr; and T, 220°C

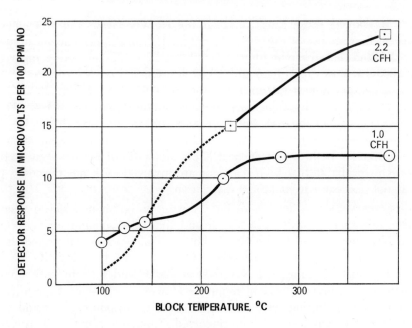

Figure 6. Effect of block temperature on response
$NH_3/NO = 2$

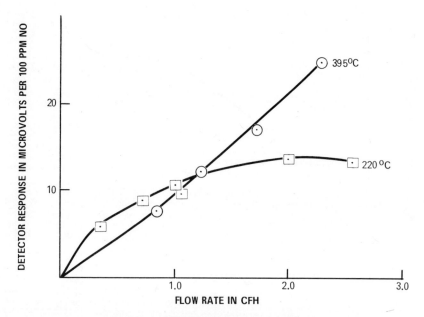

Figure 7. Effect of flow rate on response
$NH_3/NO = 2$

after prolonged usage under such reducing conditions; this was not studied.

Another possible reaction is the formation of ammonium nitrate which decomposes to $N_2O + 2\ H_2O$ at about 220°C. This may well be an intermediate, and it might cause catalyst deactivation at low temperatures or prolonged usage by filling pore space with NH_4NO_3. There was no direct evidence of this although no extended life tests were made below 220°C.

Response to Oxides of Nitrogen Other than Nitric Oxide. Although NO is normally the oxide of nitrogen in combustion products, exhaust aging at ambient temperature forms other products such as NO_2, N_2O_4, N_2O_5, and HNO_3. Catalytic muffler treatment of exhaust may produce N_2O.

Studies of gaseous HNO_3 and NO_2–N_2O_4 demonstrated essentially the same response on a nitrogen basis as for NO at 375°C and 1.0 ft³/hr. They probably decompose during preheating before contact with the detector since they can decompose in air at 398°C to produce NO as follows:

$$NO_2 \rightleftarrows N_2O_4 \rightarrow 3\ NO + 3/2\ O_2 \tag{8}$$

$$2\ HNO_3 \rightarrow 2\ NO + 3/2\ O_2 + H_2O \tag{9}$$

In the absence of a standard source of NO_x, nitric acid vapor appears useful as a secondary standard since data on the HNO_3 content of saturated nitric acid as a function of concentration and temperature are available (3). However, in order to avoid high values, the acid should be boiled to remove the dissolved NO_2 or NO usually present (as indicated by brown coloration).

Nitrous oxide was tested with and without ammonia present at concentration levels in excess of 1000 ppm at 400°C and 1 ft³/hr. There was no detectable response although N_2O could theoretically react exothermally at 1 atm and 25°C as follows:

$$N_2O \rightarrow N_2 + 1/2\ O_2 + 19.6\ kcal \tag{10}$$

$$N_2O + 2/3\ NH_3 \rightarrow 1\text{-}1/3\ N_2 + H_2O + 66\ kcal \tag{11}$$

Lack of detectability indicates that these reactions would not interfere with NO_x analysis assuming $x \geq 1$. It also suggests that N_2O is not an intermediate, but it may be an end product in the reaction of NH_3 with NO.

Possible Interferences. Reactive hydrocarbons were studied previously for this catalyst system, and they would certainly be expected to interfere seriously with analysis of most vehicle exhaust. However, these

Table I. Tests on Combustion Products of Light Gases

Source	Average Instrument Reading, ppm NO	Oxygen, %	Reading Corrected for Excess Oxygen[a], ppm NO	Handbook Flame Temperature Value, °C
Flue gas from water heater using natural gas	100	16	400	1875
Propane flame exhaust	440	5	590	1925
Hydrogen–oxygen flame exhaust	950	6	1350	2500

[a] Relative to stoichiometric.

compounds may be removed by catalytic preoxidation at 395°C before introduction of NH_3 without changing the NO_x content.

Carbon monoxide and hydrogen are also present in vehicle exhaust. Vanadia catalyst up to 400°C was inactive for their oxidation, and their presence in the absence of ammonia produced negligible response. Tests of response to 540 ppm NO with 2/1 NH_3/NO at 395°C and 1.0 ft³/hr revealed no detectable effect of CO or H_2 at levels up to 1%.

Sulfur dioxide may interfere in testing combustion gas from high sulfur fuel. Tests were therefore conducted at SO_2 levels of up to 0.5%

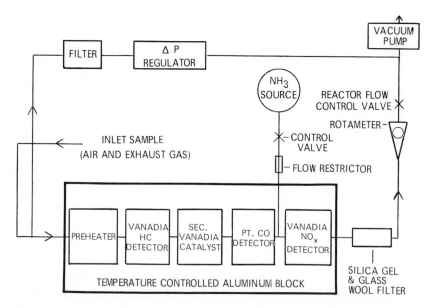

Figure 8. Flow system for simultaneous measurement of HC, CO, and NO_x in vehicle exhaust

without significant effect being detected at 395°C and 1 ft³/hr although promoted vanadia catalysts are used for SO_2 oxidation at about 500°C.

Applications

Since it was apparent that the method could be applied directly to measurement of NO_x in combustion products low in oxidizable hydrocarbons—such as combustion products from methane, carbon monoxide, and hydrogen—tests were made on flue gases from such flame sources. The findings are tabulated in Table I. When corrected for excess oxygen, the NO_x values were consistent with published flame temperatures. As expected, the hot oxy–hydrogen flame produced the highest NO_x values.

Table II. Results of Vehicle Exhaust Tests for Three-Component Thermocatalytic Analyzer

Vehicle[a]	Speed, rpm	Load	CO, %	HC,[b] ppm	NOx, ppm
1966 Valiant, V8 engine	600	none	3.0	250	200
	1200	none	1.0	200	500
	2000	none	0.4	150	1200
	2000	50 hp	0.7	160	2500
1969 Plymouth Fury III	625	none	5.2	400	100
	1000	none	0.6	140	500
	1500	none	0.6	120	700
	2000	none	0.5	120	1000

[a] After warmup.
[b] Standardized for best agreement with nondispersive infrared method (ppm as hexane).

Analysis of effluent gases from nitric acid plants that contain NO_x, O_2, H_2O, and N_2 should not present any problem, and the reaction might also be useful for NO_x control processes since an ammonia source is normally available in such plants. Analysis capability for ammonia was suggested by the data in Figure 3, and this was confirmed by studies with constant NO addition.

For application to analysis of vehicle exhaust, the system shown in Figure 8 was used. Reasonable results were obtained (Table II).

Literature Cited

1. Innes, W. B., *Environ. Sci. Technol.* (1972) **6**, 710.
2. Shelef, M., Proc. Nat. Symp. Heterogeneous Catal. Control Air Pollut., 1st, Philadelphia, September, 1968.
3. "Chemical Engineers Handbook," McGraw-Hill, New York, 1941.

RECEIVED May 28, 1974.

3

The Relative Resistance of Noble Metal Catalysts to Thermal Deactivation

G. R. LESTER, J. F. BRENNAN, and JAMES HOEKSTRA

Universal Oil Products Co., Corporate Research Center, Des Plaines, Ill. 60016

Fresh and thermally aged catalysts containing mixtures of platinum and palladium were laboratory tested for the oxidation of carbon monoxide, propane, and propylene. For both monolithic and particulate catalysts, resistance to thermal deactivation was optimum when palladium content was 80%. Full-scale vehicle tests confirmed these findings. Catalysts of this composition were developed which, on the basis of durability tests at Universal Oil Products and General Motors, appeared capable of meeting the 1977 Federal Emissions Standards with as little as 0.56 g noble metal per vehicle. The catalyst support was thermally-stabilized, low density particulate.

Platinum and palladium are the more abundant of the noble metals, and they are active for the oxidation of hydrocarbons and carbon monoxide. However, many of their characteristics as oxidation catalysts differ considerably, and these differences must be carefully considered in developing catalysts optimized to control effectively the hydrocarbon and carbon monoxide emissions of any particular vehicle. Among the important characteristics which should be considered are: (a) their relative effectiveness for converting hydrocarbons and carbon monoxide at elevated temperatures, (b) their relative low temperature effectiveness, (c) their relative resistance to the catalyst poisons potentially present in the exhaust, and (d) their relative resistance to thermally-induced deactivation.

Knowledge of these individual characteristics, and of any interactions among them, is certainly needed in order to choose the best catalyst for a particular automotive system. The system is defined by the concentrations of hydrocarbons and CO in the engine exhaust in the various engine operating modes, the mass of substrate and converter, normal and accidental concentrations of possible catalyst poisons in the engine exhaust, the operating temperatures in various modes during normal operation, and possible temperature excursions during abnormal engine operation.

The literature is limited. It is generally agreed that platinum is superior to palladium for the oxidation of paraffinic hydrocarbons, whereas palladium is preferred for the oxidation of carbon monoxide and probably unsaturated hydrocarbons also. However, at very low concentrations of active components, palladium is probably more effective than platinum for total hydrocarbon and carbon monoxide emission control, at the same weight-percent of noble metal, and this difference was more definite for CO alone; only fresh catalysts were used in this study by Barnes and Klimisch (1).

Palladium appears more susceptible than platinum to poisoning by tetraethyllead motor mix (2). Although the effect was much smaller, palladium was also slightly more susceptible than platinum to additives containing Ca, Zn, and P in the conventional lubricant used. However, phosphorus in a fuel additive affected a platinum catalyst more adversely than a palladium–base metal mixed catalyst (3).

Very little has been published about the relative low temperature effectiveness of the two metals and their relative resistance to degradation of activity after exposure to elevated temperatures. This is an investigation of these properties of platinum–palladium mixtures.

Laboratory tests were made of catalysts prepared on spherical alumina supports and on monolithic catalysts. The spherical catalysts were prepared by a proprietary technique, and were protected against shrinkage by use of a chemical stabilizer. The monoliths were wash-coated with alumina prior to impregnation with the metals. The catalysts were tested fresh and also after thermal treatment (10 hrs at 1094°C in a perfluent atmosphere consisting of 10% H_2O in air).

In the laboratory tests, two gas mixtures at two space velocities were used. Each gas mixture contained 1% CO, 250 ppm (molar) hydrocarbon, 2.5% O_2, and 10% H_2O; the remainder was N_2. The hydrocarbon was propane or propylene. The flow rate was 5000 cm^3/min at standard temperature and pressure (STP). With a catalyst bed of 20 cm^3 (7/8 × 2 in.), the gas hourly space volume (GHSV) at STP was 15,000/hr; with a bed of 2 cm^3 (7/8 × 1/4 in.), it was 150,000/hr. The test consisted of lining out the system at 538°C inlet, and following the temperature–conversion profiles while the reactor cooled until the conversions of both HC and CO were below 25%. Lined-out conversions were then obtained at 288°C and 538°C.

Although the same basic curves of performance vs. noble metal composition were obtained for both spherical and monolithic catalyst at each

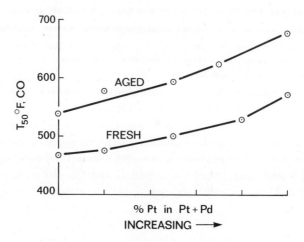

Figure 1. Inlet temperature of 50% CO conversion vs. Pt content of Pt–Pd mixtures
Total noble metals, 0.83 g/ft^3 (0.48 mg/in.^{3})

level of total noble metal loading, the effect of noble metal composition was small at the higher loadings. A typical plot for the lowest loading used (equivalent to about 100 ppm by weight of noble metal on the spherical catalyst) is given in Figure 1. CO conversion, as measured by the inlet temperature required for 50% conversion at 150,000/hr GHSV at STP, was poorer for both the fresh and the thermally aged catalysts as the proportion of platinum in the mixture increased. Significantly, the high-palladium catalysts actually performed better after thermal exposure than did the fresh high-platinum catalysts. The curves for propylene were very similar to those for CO. The curves for propane were nearly flat, but performance of both fresh and thermally-aged catalysts was slightly better with increasing platinum content up to 50%.

Zero-mile vehicle tests of the fresh catalysts also demonstrated better overall performance as palladium content increased. In fact, the zero-mile emissions were surprisingly low at levels of noble metals as low as 100 ppm in which the total content of noble metals per vehicle was as low as 0.125 gram. There was concern about the poison resistance of palladium-only catalysts. Since the laboratory findings indicated little improvement in performance after thermal aging with increasing palladium content for mixtures with more than 80% palladium, two spherical catalysts with 80% palladium content and total noble metal loadings of 0.56 g and 0.125 g per vehicle (450 and 100 ppm respectively), were given catalyst durability tests at Universal Oil Products Co. (UOP).

These durability tests define the effect of catalyst composition on catalyst durability without the confounding effect of changes in the emissions characteristic of the aging vehicle. The catalyst was therefore installed in one vehicle, and mileage accumulation was begun. About half the mileage was accumulated on a chassis dynamometer with a 4-min, 13-mode cycle including speeds above 60 mph, and the remainder was accumulated on the road with a 48-mile lap including suburban streets and expressways. The fuel was Indolene which contains 0.03 g Pb/gal, 0.01 g P/gal, and 0.01 wt % sulfur. Each 4000 miles, the converter was removed for testing on a slave vehicle which was used only for catalyst testing and which never accumulated more than 10 miles a day; during the three months of a 50,000-mile test, the slave vehicle accumulated less than 1000 miles.

In Figure 2 are plotted the least squares of the hydrocarbon emissions from the 4000–50,000-mile data for the two noble metal loadings. Baseline emissions for the slave car during the test were 1.0 g HC/mile and 15 g CO/mile; hence, overall hydrocarbon efficiency at 50,000 miles was about 81% for the catalyst with 450 ppm noble metals (PZ-236) and about 58% for the catalyst with 100 ppm (PZ-247). The PZ-236 catalyst formulation is thus a reasonable candidate to meet the 1977 statutory

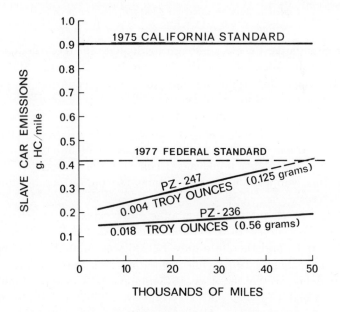

Figure 2. Hydrocarbon emissions: effect of noble metal content on durability performance of thermally optimized noble metal mixture

Emission tests on slave car (350 cid, air pump added)

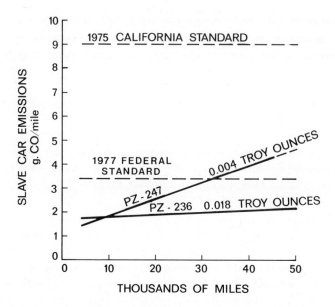

Figure 3. Carbon monoxide emissions: effect of noble metal content on durability performance of thermally optimized noble metal mixture

Emission tests on slave car (350 cid, air pump added)

Federal Standards; PZ-247, although inadequate for these Federal Standards, might be adequate for the interim 1975 California Standards of 0.9 g HC/mile. Corresponding plots for carbon monoxide are presented in Figure 3; conclusions were similar.

These tests were designed to determine the effects of various catalyst compositions and of total catalyst concentration. However, comparison of the performance of these catalysts with that of others with different composition and higher noble metal concentrations suggested that catalysts of this composition merited submission for testing by automobile manufacturers.

At the time, automobile manufacturers expected to have to meet the original statutory 1975 Federal Standards of 0.41 g HC/mile and 3.4 g CO/mile, and therefore only PZ-236 was submitted for their consideration. In March 1973, while the automakers were seeking (successfully) a delay of the statutory 1975 Standards, General Motors Corp. (GM) testified (4) that only six vehicles in their corporate fleet had accumulated 50,000 miles. All six used UOP catalysts on the same low density support. Five of the six catalysts contained 2.24 g platinum per vehicle. The sixth, PZ-236, had 25% of that amount, but in a thermally-optimized, platinum–palladium mixture. As the automaker indicated, this catalyst did not meet the Environmental Protection Agency (EPA) requirements that emissions be within the limits of 0.41 and 3.4 g/mile at both 4000 and 50,000 miles, but the failure (Figure 4) was in hydrocarbon emissions at 4000 rather than at 50,000 miles (5). Thus there was an apparent improvement in the net emissions of the system over 50,000 miles, which may reflect a change in the raw engine emissions during the test. The 4000-mile engine emissions were about 1.8 g HC/mile and 30 g CO/mile, but no data are available on the 50,000-mile raw emissions. The circle at 50,000 miles indicates the average of the two tests performed on that vehicle at the EPA laboratories. The CO emissions (Figure 5) were acceptable at both 4000 and 50,000 miles.

It would be improper to suggest that these results are the only ones obtained with PZ-236 in that durability fleet. However, the other tests were discontinued for reasons unrelated to the efficiency of PZ-236 as a catalyst. In one test, the converter was empty at 36,000 miles because of a hole in the outlet screen, while engine failures were cited as the cause for discontinuation of the other tests. In all the tests, however, the emissions were within the original 1975, and now 1977, standards at the time of test termination.

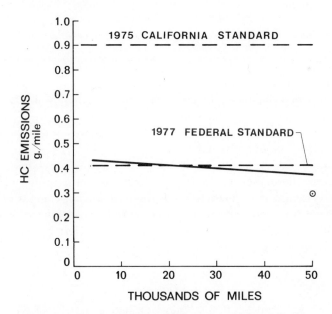

Figure 4. Hydrocarbon emissions of car from GM durability fleet with HN-1682 (PZ-236) catalyst containing 0.56 g noble metal

Buick #1258, 350-4V, 260-in.³ underfloor converter; deterioration factor, 0.85

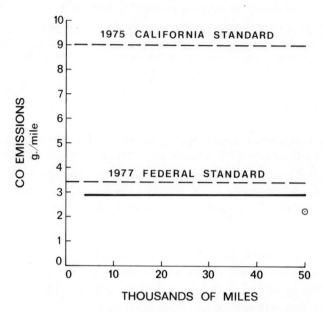

Figure 5. Carbon monoxide emissions of car from GM durability fleet with HN-1682 (PZ-236) catalyst containing 0.56 g noble metal

Buick #1258, 350-4V, 260-in.^{3} underfloor converter; deterioration factor, 1.01

In conclusion, it appears that catalysts containing a noble metal mixture, properly balanced to give maximum thermal stability and an appropriate degree of protection against poisons, can meet the 1975 California Standards or the 1977 Federal Standards with surprisingly small quantities of noble metal.

It should not, however, be concluded that this particular mixture, optimized for thermal exposure, is necessarily the optimum for all vehicle converter systems. It is our belief that the noble metal composition must be optimized for each particular system, with due consideration of inlet temperatures, converter mass, concentrations of catalyst poisons, and the likely thermal experience of the catalyst bed.

Literature Cited

1. Barnes, G. J., Klimisch, R. L., "Initial Oxidation Activity of Noble Metal Automotive Exhaust Catalyst," Soc. Automot. Eng., Auto. Eng. Mtg., Detroit, May, 1973, paper **730570**.
2. Gallopoulos, N. E., Summers, J. C., Klimisch, R. L., "Effects of Engine Oil Composition on the Activity of Exhaust Emissions Oxidation Catalyst," Soc. Automot. Eng., Auto. Eng. Mtg., Detroit, May, 1973, paper **730598**.

3. Giacomazzi, R. A., Homfeld, M. F., "The Effect of Lead, Sulfur, and Phosphorus on the Deterioration of Two Oxidizing Bead-Type Catalysts," Soc. Automot. Eng., Auto. Eng. Mtg., Detroit, May, 1973, paper **730595**.
4. "General Motors Request for Suspension of 1975 Federal Emissions Standards," submitted to Environmental Protection Agency March 5, 1973, vol. I, sec. 6, pp. 3–12.
5. "General Motors Request for Suspension of 1976 Federal Emissions Standards," submitted to Environmental Protection Agency June 20, 1973, vol. III, sec. 31, pp. 5–9.

RECEIVED September 19, 1974.

though
4

Variation of Selectivity with Support Chemistry in NO$_x$ Removal Catalysts

GEOFFREY E. DOLBEAR and GWAN KIM

W. R. Grace & Co., Davison Chemical Division, Columbia, Md. 21044

Ammonia selectivity of platinum and platinum–nickel catalysts for NO$_x$ reduction varies with the nature of the supporting oxide. Silica, alumina, and silica–alumina supports on monolithic substrates were studied using synthetic automotive exhaust mixtures at 427°–593°C. The findings are explained by a mechanism whereby the reaction of nitric oxide with adsorbed ammonia is in competition with ammonia desorption. The ease of this desorption is affected by the chemistry of the support. Ammonia decomposition is not an important reaction on these catalysts when water vapor is present.

The catalytic removal of nitrogen oxides from automotive exhaust gases has been the subject of many studies. Catalysts containing at least 20 different metals, alone and in combination, have been tested. We found that the support used in catalyst preparation is as important as the metal, particularly in catalyst selectivity toward nitrogen rather than ammonia in strongly reducing streams. This paper is a report on some of the effects of support chemistry in a fairly well known system, platinum-promoted nickel (*1*). We also elucidate the pathways of ammonia removal in this system.

Related work on supports was performed by Balgord and Wang (*2*); they explored the reaction NO + CO in dry systems and found that the support was very important in NO disappearance activity. In our work with simulated exhaust streams containing hydrogen and water, support chemistry had more effect on selectivity than on activity.

We compared monolithic catalysts by keeping constant support surface areas, preparative techniques, and active compositions while varying only the support chemistry. Support pore volume could not be held

Table I. Properties of Catalyst Supports

Material	Calcination Temperature, °C	N_2 Surface Area, m^2/g	N_2 Pore Volume, cm^3/g
Silica, Davison grade 81	927	167	0.33
Silica, Davison grade 952	871	158	0.95
Alumina	899	155	1.14
Silica–alumina (cracking catalyst)	927	168	0.40
Mullite	954	170	0.94

constant, so we compared two silicas of different porosity and found selectivity differences. Catalysts were tested using gas compositions and reaction temperatures selected after examination of second-by-second temperature and gas compositions recorded during the Environmental Protection Agency test (3). Under these conditions with platinum and nickel catalysts, nitric oxide disappearance was virtually complete, with typical conversions being above 90% and even the worst catalyst giving over 70% removal. The emphasis in the interpretation, then, is on selectivity toward nitrogen.

Experimental

The supports used were our own materials. Silica gels, grades 952 and 81, were commercially available powders from the Davison Chemical Division of W. R. Grace & Co. The silica–alumina from Davison was a nonzeolite-containing, semi-synthetic cracking catalyst containing 1:2 kaolin:synthetic silica–alumina (25% alumina). Both the alumina and the mullite were experimental materials. The alumina was a gel material which was primarily gamma by x-ray. The mullite was prepared by the Rundell, Kwedar, and Duecker method (4), and it contained 72% aluminum oxide. Calcination temperatures were chosen for each oxide to yield a final surface area of 160 ± 10 m^2/g. Properties of the supports are presented in Table I.

Catalysts were prepared by incipient wetness impregnations using nickel nitrate and tetraamine–platinum(II) nitrate; separate impregnations were made with air calcination at 593°C to decompose the nickel salts before platinum was added. After a second calcination, the powders were ball-milled with water to form a coating slurry which was used to coat Poramic (from W. R. Grace) monoliths; slurry compositions were adjusted to give 10% solids pickup. The metal levels did not depend on support density. See Table II for final compositions.

Activities were measured in all-glass, vertical, tubular reactors with Alundum balls in the preheat section. A metal-free reactor was used because our earlier work had shown that type 316 stainless steel reactors converted 20–30% of the NO to NH_3 in the absence of a catalyst; there was no background conversion with glass.

Gas compositions (Table III) were representative of those observed during automotive emissions testing. Gas flow (10 l/min) and sample

Table II. Composition of Catalysts

Catalyst	Support	Pt, %	Ni, %	Support, %
1	Silica, grade 81	0.1	0	9.9
2	Silica, grade 81	0.1	0.5	9.4
3	Silica, grade 81	0.1	1.3	8.6
4	Alumina	0.1	0	9.9
5	Alumina	0.1	0.5	9.4
6	Alumina	0.1	1.3	8.6
7	Silica–alumina	0.1	0	9.9
8	Silica–alumina	0.1	0.5	9.4
9	Silica–alumina	0.1	1.3	8.6
10	Mullite	0.1	0.5	9.4
11	Silica, grade 952	0.1	0.5	9.4

size (monolith 1 in. in diameter and 0.5 in. long) gave exhaust gas space velocities (EGSV) of 100,000/hr (adjusted to standard temperature and pressure). These were also typical of automotive conditions; an EGSV of 150,000/hr is calculated for a 350-cid engine operating at intermediate speeds (1500 rpm) with the exhaust passing over a pair of 75-in.[3] monolithic catalysts.

Table III. Composition of Simulated Exhaust Gas

Gas	Content, %
CO	3.0
H_2	1.5
NO	0.10 (1000 ppm)
C_3H_8	0.025 (250 ppm)
O_2	0.20
H_2O	10
CO_2	10
N_2	balance

Gas mixtures were prepared by mixing reagent gases from cylinders; deionized water was vaporized in the heated stream outside the reactor. After passing over the reactor, the gases were cooled to remove the water, and a portion was diverted to an analytical train consisting of nondispersive IR analysis for CO and fuel cell analysis for NO (both Envirometrics and Dynasciences units were used). Instruments were standardized with gas mixtures from Scott Research Laboratories. Ammonia was determined in two ways. An Orion ammonia electrode was used to analyze samples scrubbed into dilute hydrochloric acid; or oxygen was added to the reaction effluent and the mixture was passed over platinum on alumina at 604°C to reoxidize the ammonia to nitric oxide with net conversions to nitrogen being read directly. The scrubbing method was used to calibrate the oxidation method.

Ammonia decomposition and nitric oxide–ammonia activities were measured similarly, with 1000 ppm of the desired components and 10% water vapor (remainder N_2) in the feeds.

Results and Discussion

All the catalysts were active for NO removal (Table IV). In the absence of nickel, the most important product was ammonia with very little difference attributable to the support. The importance of nickel had been reported (1, 5, 6). The effect of a given increment of nickel was markedly dependent on the support; silica gave the most response, alumina the least (see Figure 1). Changing the pore volume of the silica support also strongly affected the response to nickel. The changed pore geometry probably caused diffusion differences for intermediate ammonia which would stay longer in the region where it could be consumed in later reactions (see below). The change in pore volume may change nickel dispersion which could also cause the noted effects.

At first we interpreted the effects of the support in terms of acidity since a more acid surface would tend to trap ammonia for further reaction. Silica aluminas, including zeolites, were more selective than alumina, and the effects were also achieved by using other acidic supports like silica–magnesia and silica–zirconia cracking catalysts. Catalysts supported on synthetic spinel ($MgAl_2O_4$), on the other hand, had poor selectivity. The findings with silica jeopardized this interpretation until we realized that nickel oxide, like magnesium oxide for example, would enhance silica acidity. This hypothesis also explains the high sensitivity of silica to nickel since two important variables (presence of nickel and surface acidity) are being increased simultaneously. On the other hand, one would expect alumina to lose both active nickel and acidity by reaction with nickel oxide to form a surface spinel; this is consistent with the observations.

Table IV. Nitric Oxide Removal Activities

Catalyst	482°C		538°C		593°C	
	Total[a]	Net[b]	Total	Net	Total	Net
1	82	22	87	22	88	23
2	98	46	99	64	99	84
3	99	56	99	75	99	91
4	91	26	90	23	87	24
5	95	21	95	21	94	32
6	98	30	98	42	98	64
7	71	15	78	16	83	21
8	91	20	92	33	94	54
9	95	43	96	60	97	79
10	95	18	95	40	96	66
11	95	32	96	54	97	78

[a] Disappearance of NO (%).
[b] Disappearance of NO (%) after reoxidation of NH_3 to NO.

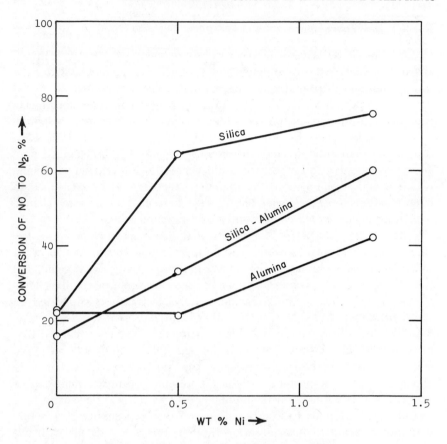

Figure 1. Conversion of NO to N_2 at 538°C

We explored the mechanistic implications of the apparent importance of acidity to ammonia selectivity, beginning with the assumption that ammonia is an intermediate in the reduction of nitric oxide over precious metal catalysts. (Attempts to explain our observations on the alternative assumption that chemisorbed isocyanate is the active intermediate (7) were fruitless since nickel should not change isocyanate's behavior. However, an acidic surface would probably lower isocyanate stability.) Since the total conversion of NO increased with nickel addition, it is likely that some conversion takes place on the nickel, probably by the reaction of NO with hydrogen spilled over from the platinum. Disappearance of this ammonia could occur by ammonia decomposition (Reaction 1) as hy-

$$2NH_3 = N_2 + 3H_2 \tag{1}$$

pothesized by Klimisch and Taylor (1), or by reaction with NO to form either nitrogen (Reaction 2) or nitrous oxide (Reaction 3) (8). In order

Table V. Ammonia Disappearance Activities[a]

Catalyst	Ammonia Decomposition NH_3 Removal[b], %	Ammonia–Nitric Oxide Reaction	
		NH_3 Removal[c], %	NO Removal, %
1	0	35	67
2	15	46	90
3	25	67	96
4	41	41	73
5	3	61	96
6	36	65	100
7	9	31	67
8	5	51	88
9	28	54	100
10	13	50	91
11	18	56	97

[a] Measured at 593°C and 100,000/hr EGSV.
[b] NH_3, 1000 ppm; H_2O, 10%; and N_2 inlet, 89.9%.
[c] NH_3, 1000 ppm; NO, 1000 ppm; H_2O, 10%; and N_2 inlet, 89.8%.

$$2NH_3 + 3NO = 5/2N_2 + 3H_2O \quad (2)$$
$$2NH_3 + 8NO = 5N_2O + 3H_2O \quad (3)$$

to differentiate between these possibilities, both the ammonia decomposition and the NO–NH_3 activities of our catalysts were measured (Table V). Ammonia decomposition in water vapor-containing streams was not a facile reaction, with the highest measured activity appearing on an otherwise poor catalyst (no. 4: Pt on Al_2O_3). The NO–NH_3 reaction, however, was much more rapid and was limited in several instances by the stoichiometry of the feed gases. The stoichiometry was measured for a typical catalyst (no. 8: 0.5% Ni, 0.1% Pt, silica–alumina), and the ratio of moles NO to moles NH_3 reacted was calculated (Table VI). Only at the highest temperature did the value approach the 1.5 required by Reaction 2; the values increased as the temperature dropped which indi-

Table VI. Ammonia–Nitric Oxide Reaction Stoichiometry[a]

Bed Temperature, °C	Moles NO : Moles NH_3 Reacted
482	1.91
538	1.59
593	1.63
649	1.60

[a] Catalyst 8; NO, 1000 ppm; NH_3, 1000 ppm; H_2O, 10%; N_2, 89.8%; 100,000/hr EGSV.

cates that nitrous oxide was also being produced by Reaction 3. One of these product mixtures contained over 200 ppm N_2O by gas chromatographic separation. [Nitrous oxide was also reported by Shelef and Otto (8).] Only if the value of the mole ratio were less than 1.5 could we conclude that ammonia decomposition, Reaction 1, makes an important contribution to the system.

We also observed, with nickel-promoted palladium catalysts not reported on here, that ammonia decomposition was poisoned by CO as well as water. If the same occurs on platinum, then the high NH_3 formation at high CO levels could result from competition of NO and CO for the active sites. We believe that this happens with our platinum–nickel catalysts although measurements of this inhibition were obscured by reactions which consumed CO.

Literature Cited

1. Klimisch, R. L., Taylor, K. C., *Environ. Sci. Technol.* (1973) **7** (2), 127.
2. Balgord, W. D., Wang, K.-W. K., Proc. Int. Clean Air Congr., 2nd, Washington, December, 1970, pp. 818–826.
3. *Fed. Regist.* (July 15, 1970) **35** (136), part II.
4. Rundell, C. A., Kwedar, J. A., Duecker, H. C., W. R. Grace & Co., U.S. Patent **3,533,738** (1970).
5. Bernstein, L. S., *et al.*, Soc. Automot. Eng., Automot. Eng. Congr., Detroit, January, 1971, paper **710014**.
6. Bernstein, L. S., *et al.*, Soc. Automot. Eng. Auto. Eng. Mtg., Detroit, May, 1973, paper **730567**.
7. Unland, M. L., *J. Phys. Chem.* (1973) **77** (16), 1952.
8. Shelef, M., Otto, K., *J. Catal.* (1968) **10**, 408.

RECEIVED May 28, 1974.

5

Thermodynamic Interaction between Transition Metals and Simulated Auto Exhaust

HOWARD D. SIMPSON

Union Oil Co. of California, Union Research Center, Brea, Calif. 92621

> A computer study of the equilibrium concentrations of the major constituents in auto exhaust, and of the equilibrium relations of various transition metals with these constituents, provides some interesting insights into the state of chemical combination expected for metals used as NO_x catalysts. The state of chemical combination of most metals is not affected appreciably by changes in the air/fuel ratio in the normal range. The sulfates of many metals can form in auto exhaust under nearly all normal conditions. Sulfides and oxides can form in much of the normal range. Thus, sulfating, sulfiding, and oxidation can all contribute significantly to catalyst deactivation.

The elimination of NO_x from automobile exhaust is a major objective of industrial scientific research. One method of eliminating this pollutant is catalysis.

The fundamental purpose of using a catalyst is to cause slow chemical reactions to reach equilibrium rapidly. The effectiveness of a catalyst can be assessed by measuring the extent to which equilibrium has been approached in the system in which it is being used. In studies connected with any catalytic process, it is therefore extremely useful to determine in advance, if possible, the equilibrium concentrations of the various reactants and products. In catalyst design and development, it is also very useful to have an idea about the states of chemical combination expected for the candidate catalyst metals. Thus, knowledge about the equilibrium state of metals as well as about the concentrations of the appropriate substances in the system of interest is necessary. These objectives were pursued in the Union Oil Co.'s NO_x catalyst program.

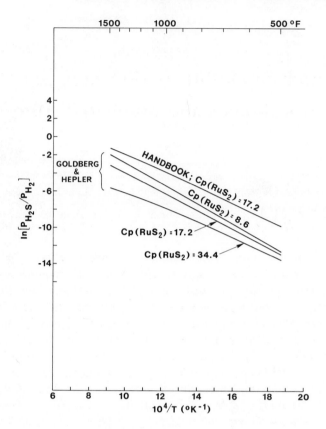

Figure 1. Comparative values of $\ln[P_{H_2S}/P_{H_2}]$ vs. $10^4/T$ for $RuS_2 + 2H_2 \to Ru + 2H_2S$
Data from "Handbook of Chemistry and Physics" (1) and from Goldberg and Hepler (2)

Calculation of Equilibrium Constants

This work is based on equilibrium constants for four categories of chemical reactions:

(a) Reactions expected to occur over a NO_x catalyst between some of the constituents in synthetic automobile exhaust gas,

(b) Reactions between transition metal sulfides and hydrogen,

(c) Reactions between transition metal oxides and hydrogen, and

(d) The formation of the sulfates of lead, copper, and nickel from the metals and metal sulfides.

Almost all the equilibrium constants were calculated by our program based on the equation

$$\ln K_T = -\frac{\Delta F_T}{RT} = -\frac{\Delta H_o}{RT} + \frac{\Delta(na)\ln T}{R} +$$
$$\frac{\Delta(nb)T}{2R} + \frac{\Delta(nc)T^2}{6R} + C \qquad (1)$$

where K_T is the equilibrium constant at absolute temperature T; ΔF_T is the free energy at that temperature; $\Delta(na)$, $\Delta(nb)$, and $\Delta(nc)$ are the molar summations of the heat capacity equation coefficients for the products and reactants involved; ΔH_o and C are integration constants which permit the equation to be based on any desired set of reference ΔH and ΔF values; and R is the gas constant.

Data were obtained as follows: the reference ΔH and ΔF values for gaseous reactions and for most of the reactions involving base metals (1),

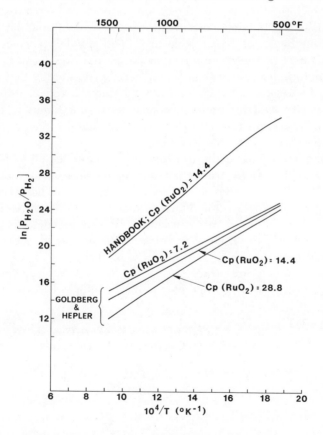

Figure 2. Comparative values of $\ln[P_{H_2O}/P_{H_2}]$ vs. $10^4/T$ for $RuO_2 + 2H_2 \rightarrow Ru + 2H_2O$

Data from "Handbook of Chemistry and Physics" (1) and from Goldberg and Hepler (2)

for reactions involving noble metals (1, 2); and heat capacity data (3, 4, 5).

The accuracy of the calculated equilibrium constants for gaseous reactions was good because of the high accuracy of experimentally determined thermodynamic data for gases. As a result, the values obtained for reactions such as $NO \rightarrow \frac{1}{2} N_2 + O_2$, $NO + CO \rightarrow \frac{1}{2} N_2 + CO_2$, and $CO + H_2O \rightarrow CO_2 + H_2$ agreed well with the values reported elsewhere (6, 7).

Equilibrium constants obtained for reactions involving metals were much less accurate because the experimental data on systems involving solids vary greatly. It is difficult to obtain good thermodynamic data on solids because of nonstoichiometric compositions, surface phases, and partial phase transformations. For example, Nunez et al. (8) found that the heat of formation of CuO depended on mode of preparation, state of subdivision, and previous heat treatment. Annealed, granular CuO was more stable by about 1 kcal/mole than finely divided powder.

The effect of these variations on the partial pressure ratios derived from equilibrium constants can be realized by considering the interaction between hydrogen and RuS_2 or RuO_2 (Figures 1 and 2, respectively). Heat capacity data for these substances were not found so they were estimated by Kopp's rule (9). The effect of any error incurred in estimating the heat capacities is reflected by the difference between the curve obtained by using the true estimated (intermediate) value of Cp and those obtained by using half and twice that value (see the sets labeled "Goldberg & Hepler" in Figures 1 and 2). The other type of uncertainty illustrated in the figures is that due to differences in the ΔH and ΔF values from different sources (cf. the intermediate curve in the Goldberg set and the Handbook curve. Both sources of uncertainty are substantial, but those due to differences in the ΔH and ΔF values are probably more serious. Uncertainties in the natural logarithm of the partial pressure ratio attributable to these differences are about 4–10

Table I. Simulated Auto Exhaust Used in Union Oil NO_x Catalyst Evaluation Unit

Component	Content, mole %
H_2	0.33
O_2	0.35
H_2O	10.00
CO	2.00
CO_2	13.00
C_3H_6 (or C_3H_8)	0.10
NO	0.08
N_2	74.14

Table II. Sets of Independent Chemical Reactions Assumed to Occur Among the Various Components of the Simulated Exhaust Gas in Table I

Model	Chemical Reactions
A[a]	$CO + H_2O \rightarrow CO_2 + H_2$
B[a]	$H_2 + 1/2 O_2 \rightarrow H_2O$ $CO + H_2O \rightarrow CO_2 + H_2$
I[b]	$H_2 + 1/2 O_2 \rightarrow H_2O$ $CO + H_2O \rightarrow CO_2 + H_2$ $C_3H_6 + 3O_2 \rightarrow 3CO + 3H_2O$
I-A[b]	same as I except 3rd reaction is replaced by: $C_3H_6 + 9/2 O_2 \rightarrow 3CO_2 + 3H_2O$
II	same as I with addition of: $NO + CO \rightarrow 1/2 N_2 + CO_2$
III	same as I with addition of: $2NO + 2H_2 \rightarrow N_2 + 2H_2O$
IV	same as I with addition of: $2NO + 5H_2 \rightarrow 2NH_3 + 2H_2O$
V[c]	same as I with addition of: $2NH_3 \rightarrow N_2 + 3H_2$

[a] C_3H_6, NO, N_2, and O_2 were treated as inerts in Model A, and all except O_2 were treated as inerts in Model B.
[b] NO and N_2 were treated as inerts in these runs.
[c] 50% of the NO (0.04 mole %) was replaced by N_2 and 50% by NH_3 in the initial gas composition for this run to simulate the operation of a dual-function catalyst.

units. This represents a factor of about 10^2–10^5 in the partial pressure ratio. The data from Goldberg and Hepler (2) indicate that RuS_2 and RuO_2 are more stable than would be surmised from the Handbook data (1).

Although the uncertainty in dealing with the thermodynamics of noble metals is enormous, the data can still be of value in speculating about the state of chemical combination of these metals in simulated auto exhaust (see below). The Goldberg and Helpler data on noble metals were used throughout this work because it was felt that these were the most dependable values available.

Compounds for which Cp values were estimated by Kopp's rule were IrS_2, PdS, PtO, RuO_2, and RuS_2. Sources of other data were as follows: equilibrium data for the interaction of NiO with hydrogen (10), data for the interaction of MoS_2 and PbS with hydrogen (4), and the ΔH for PdS (2). No value of ΔF for PdS was available; consequently, it was assumed to equal $\Delta H_{PdS} - (\Delta H - \Delta F)_{PtS}$.

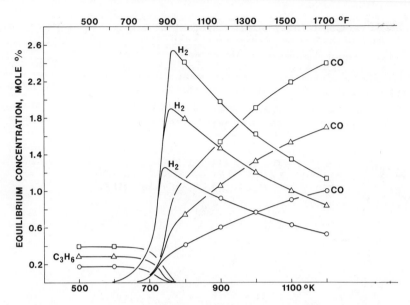

Figure 3. Calculated equilibrium concentrations of major constituents in simulated auto exhaust for Model I

Initial CO: ○, 1 mole % (A/F ratio = 14.2); △, 2 mole % (A/F = 13.9); and □, 3 mole % (A/F = 13.7)

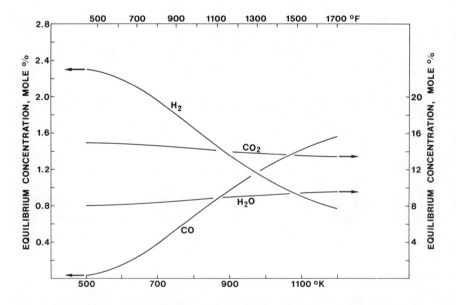

Figure 4. Calculated equilibrium concentrations of major constituents in simulated auto exhaust for Model A

Initial CO level, 2 mole %

Equilibrium Composition of Simulated Auto Exhaust

The normal composition of the simulated exhaust fed to Union Oil's bench scale NO_x catalyst evaluation unit is listed in Table I. This composition simulates the auto exhaust obtained with an air/fuel (A/F) ratio of 13.9 (*11*). At times, catalysts were also evaluated with CO = 1.00 and 3.00 mole % with A/F = 14.2 and 13.7 respectively. This range

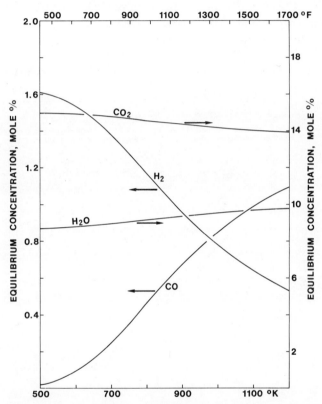

Figure 5. Calculated equilibrium concentrations of the major constituents in simulated auto exhaust for Model B

Initial CO level, 2 mole %

of CO levels represents the range of A/F ratios encountered in normal automobile operation.

Equilibrium compositions of the simulated exhaust were calculated as a function of temperature for several assumed reaction models. The calculations consisted of solving simultaneously, by the Newton iteration procedure, a set of material balance equations and chemical equilibrium constraints for the temperature and reactions involved. The reaction models considered are summarized in Table II.

All the models in which the reduction of NO was considered predicted nearly total elimination of this constituent at equilibrium at all temperatures of interest. Model V predicted that at least 95% of any ammonia formed would be decomposed at equilibrium. Model I represents fairly well the behavior of the major constituents in simulated auto exhaust under the influence of an active catalyst. This model is depicted graphically in Figure 3, for all three initial CO levels. The reactive hydrocarbon propylene disappeared between 427° and 538°C (800°–1000°F). The predicted equilibrium hydrogen concentration at 667°C (1250°F) with 1 mole % CO initially in the feed gas was 0.87%; a value of 0.94% was measured by mass spectrometry. Most surprising about

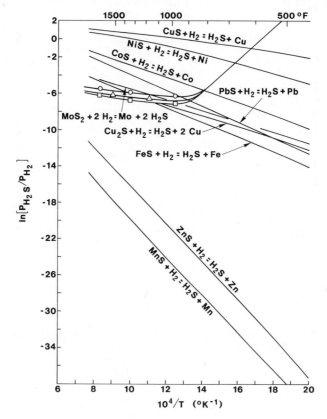

Figure 6. Superposition of system $\ln[P_{H_2S}/P_{H_2}]$ curves on base metal equilibrium desulfiding curves for simulated exhaust gas containing 20 ppm H_2S

CO: ○, 1 mole % (A/F ratio = 14.2); △, 2 mole % (A/F = 13.9); and □, 3 mole % (A/F = 13.7)

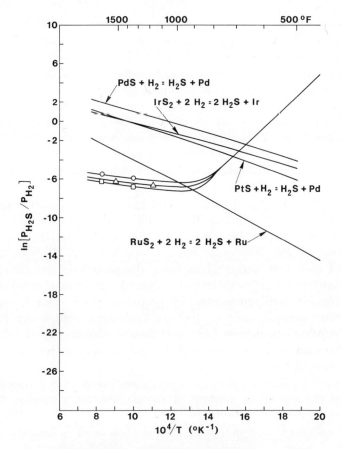

Figure 7. Superposition of system $\ln[P_{H_2S}/P_{H_2}]$ curves on noble metal equilibrium desulfiding curves for simulated exhaust gas containing 20 ppm H_2S

CO: ○, 1 mole % (A/F ratio = 14.2); △, 2 mole % (A/F = 13.9); and □, 3 mole % (A/F = 13.7)

the model was the virtual absence of hydrogen and CO below 315°C (600°F), with a resultant increase in hydrocarbon concentration. The Fisher–Tropsch synthesis was therefore incorporated in the model at low temperatures. This behavior was not predicted by Models A and B, *i.e.* those which do not include the hydrocarbon oxidation reaction. These models are depicted in Figures 4 and 5. Model I-A was inadequate since it did not predict hydrocarbon disappearance until about 760°C (1400°F) which is contrary to observation.

The general features of Model I agree with the work of Barnes, Klimisch, and Krieger (*12*). They calculated the equilibrium concentrations of a substantial number of constituents expected in auto exhaust obtained at A/F ratios of 12.1–17.1 by a free-energy minimization technique. The hydrogen concentration did not vary as abruptly in their true equilibria models as in Model I, but they, too, predicted that hydrogen concentration would be maximum around 482°C (900°F) and that hydrogen concentration would diminish greatly below this temperature. They also predicted the disappearance of hydrocarbons above 482°C. Their findings demonstrated that methane is essentially the only hydrocarbon formed by the Fisher–Tropsch synthesis in the system. Model I differs in that all hydrocarbons are expressed as propylene; it was decided to account for the hydrocarbons in this way rather than to make provisions for methane formation because very little methane was ever detected by infrared (IR) analysis of our simulated exhaust after exposure to several active NO_x catalysts. In fact, it was difficult to detect the IR spectra of any familiar hydrocarbon species. On the other hand, the low level of propylene in the feed was detected quite easily. It was therefore assumed that, with the catalysts used, the propylene was converted to some distribution of hydrocarbons, alcohols, and aldehydes which might be reasonably well represented by propylene itself. In this sense, the equilibrium computations in this work represent a departure from the true equilibria calculations (12), and they emphasize that the degree to which true equilibrium is attained depends on the catalyst to which the system is subjected.

Pseudo-equilibria calculations (12) indicated that ammonia originates in the system as a result of the inability of a catalyst to effect hydrocarbon formation; under the influence of such a catalyst, the system does not attain true equilibrium. In the absence of hydrocarbon formation, the hydrogen concentration remains high throughout the temperature range, favoring the formation of ammonia at low temperatures.

Table III. Predicted Desulfiding Temperature for Various Metals in Simulated Auto Exhaust Containing 20 ppm H_2S

Metal	Desulfiding Temperature, °C
Co	427
Cu	649 (Cu_2S)
Fe	649
Ir	371
Mn	none
Mo	538
Ni	371
Pb	482
Pd	371
Pt	371
Ru	482
Zn	none

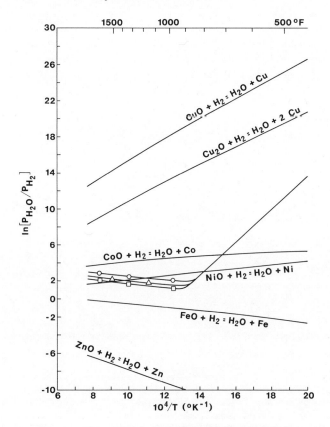

Figure 8. Superposition of system $\ln[P_{H_2O}/P_{H_2}]$ curves on base metal equilibrium deoxiding curves

CO: ○, *1 mole % (A/F ratio = 14.2);* △, *2 mole % (A/F = 13.9); and* □, *3 mole % (A/F = 13.7)*

The State of Chemical Combination of Transition Metals in Simulated Auto Exhaust

To consider the propensity for forming metal sulfides in simulated auto exhaust, the equilibrium $\ln[P_{H_2S}/P_{H_2}]$ values were plotted as a function of reciprocal temperature for the interaction of a metal sulfide with hydrogen to form the metal and H_2S. If it is assumed that all SO_2 in typical auto exhaust is converted to H_2S by a NO_x catalyst, then a fuel containing 300 ppm sulfur would yield a gas containing about 20 ppm (by volume) H_2S. This is equivalent to 0.0020 mole %. Using the hydrogen concentrations from Figure 3, the function $\ln[P_{H_2S}/P_{H_2}]$ for the system can then be expressed as $\ln[0.0020/(\text{mole \% } H_2)]$. This function is superimposed on the equilibrium curve for the metal. At temperatures where the values of $\ln[P_{H_2S}/P_{H_2}]$ are higher on the system

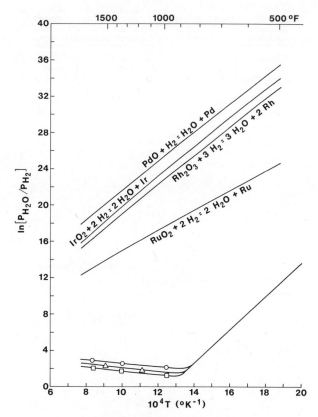

Figure 9. Superposition of system $\ln[P_{H_2O}/P_{H_2}]$ *curves on noble metal equilibrium deoxiding curves*
CO: ○, 1 mole % (A/F ratio = 14.2); △, 2 mole % (A/F = 13.9); and □, 3 mole % (A/F = 13.7)

curve than on the metal equilibrium curve, the metal will tend to sulfide. At temperatures where the reverse is true, the metal will tend to desulfide.

Such a scheme for base metals is presented in Figure 6. An analogous system for noble metals is depicted in Figure 7. These plots demonstrate that, in general, metals will tend to sulfide at low temperatures and to desulfide at high temperatures. Also, noble metals sulfide nearly as readily as base metals. With the exception of copper, iron, and molybdenum, the CO level (A/F ratio) had essentially no effect on the desulfiding temperature. The desulfiding temperatures are summarized in Table III.

Analogously, deoxiding charts can be prepared by superimposing the system $\ln[P_{H_2O}/P_{H_2}]$ function on equilibrium $\ln[P_{H_2O}/P_{H_2}]$ curves for various metals. Figure 8 depicts such a chart for base metals, Figure 9

Table IV. Predicted Deoxiding Temperatures for
Various Metals in Simulated Auto Exhaust

Metal	Deoxiding Temperature, °C
Co	371
Cu	<260
Fe	none
Mn	none
Ni	>427 and <649
Zn	none
all noble metals	<260

one for noble metals. Copper is the only base metal which should not be extensively oxidized in the system. None of the noble metals should be oxidized. The data are in Table IV. Note that nickel is predicted to reduce above 427°C and then to oxidize again above about 649°C.

Sulfating equilibrium plots are presented in Figure 10 for copper, lead, and nickel. The huge values of the equilibrium constants indicate that base metal sulfates form readily as long as any oxygen is present. There is little evidence that sulfates of noble metals are stable.

Implications for NO_x Catalysis

One of the primary requisites for a good redox catalyst is metallic character. Any material which loses this property upon exposure to automobile exhaust will not be an effective NO_x catalyst. From the above considerations, one would predict that all noble metals active for NO_x reduction will be immune from chemical poisoning by sulfur and oxygen. On the other hand, any base metal active for NO_x reduction will be chemically poisoned by sulfide or sulfate formation as long as any sulfur is present in the system. These predictions are generally confirmed by experience. Any base metal other than copper will probably be chemically poisoned by oxygen in the molecular form or derived from water.

As was demonstrated above, the uncertainty in the natural logarithms of the equilibrium partial pressure ratios describing the interaction between noble metals and H_2O or H_2S could be 4–10 units; the greatest discrepancies emanate from different ΔH and ΔF values for noble metal oxides and sulfides. Discrepancies for base metals were less severe; the maximum uncertainty in the natural logarithm of the partial pressure ratio is probably only four or five units. According to the Goldberg and Hepler data for noble metals (2), the noble metal oxides and sulfides are much more stable than is indicated by the data in the "Handbook of Chemistry and Physics" (1). Therefore, our desulfiding and deoxiding temperatures for noble metals may be high by up to about 111°C. The

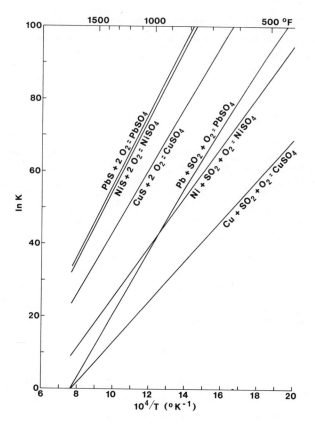

Figure 10.   ln K vs. $10^4/T$ for the formation of copper, lead, and nickel sulfates

uncertainty expected for base metals is about 56°C. Thus, noble metals may actually be reduced more easily than predicted for simulated auto exhaust so that our conclusions are on the conservative side and not invalid.

When equilibria are calculated using the gas composition in Table I and including the methanation reaction $CO + 3H_2 \rightarrow CH_4 + H_2O$ in Model I, the findings are in virtual agreement with the true equilibria data of Barnes, Klimisch, and Krieger (12). The principal difference between true equilibrium and the data depicted in Figure 3 is that the hydrogen concentration was substantially higher for the former at 260°–371°C. At true equilibrium, then, the system logarithmic partial pressure functions in Figures 6, 7, 8, and 9 would not increase as rapidly as indicated below 482°C ($10^4/T \geq 13$). The magnitudes of the differences are such that metals predicted to reduce at temperatures below 482°C in this work would be predicted to reduce at temperatures about 56°–111°C lower at true equilibrium. Thus, nickel, perhaps cobalt, and all noble metals except ruthenium would be reduced thermodynamically at all temperatures of interest if true equilibrium were attained.

This permits an interesting insight into the mode of operation of base metal NO_x catalysts promoted by noble metals. We found that nickel promoted by platinum was more active for NO_x conversion in simulated automobile exhaust than platinum alone (13). Moreover, nickel alone was quite inactive for NO_x conversion except in systems where neither water nor oxygen was present; then it was very active. It therefore seems probable that nickel alone does not cause true equilibrium to be attained in auto exhaust, presumably because of kinetic limitations, whereas platinum does so to a substantial extent. Once true equilibrium has been induced by the platinum, the nickel reduces and becomes active also; the resultant activity is greater than could be attained with either platinum or nickel alone. This concept was supported by our recent observation that the platinum did not have to be intimately associated with the nickel to cause it to become active. A nickel catalyst became active above 371°–427°C at 138,000 gaseous hourly space velocity in simulated auto exhaust when a separate platinum catalyst was placed upstream. It is possible that some additional benefit might be obtained if the two metals were intimately associated. The extent of such benefit, however, has not yet been determined.

Literature Cited

1. "Handbook of Chemistry and Physics," 48th ed., p. D-38, Chemical Rubber Co., Cleveland, 1967.
2. Goldberg, R. N., Hepler, L. G., *Chem. Rev.* (1968) **68** (2), 229.
3. Kelley, K. K., U.S. Bur. Mines Bull. (1937) **406**.
4. Kelley, K. K., U.S. Bur. Mines Bull. (1962) **601**.
5. "Metals Reference Book," 3rd ed., vol. 2, Butterworth, Washington, D.C., 1962.
6. Shelef, M., Kummer, J. T., *Chem. Eng. Prog. Symp. Ser.* (1971) **67** (115), 74.
7. "Catalyst Handbook," pp. 192–193, Springer Verlag, New York, 1970.
8. Nunez, L., Pilcher, G., Skinner, H. A., *J. Chem. Thermodyn.* (1969) **1** (1), 31.
9. Smith, J. M., "Introduction to Chemical Engineering Thermodynamics," p. 97, McGraw-Hill, New York, Toronto, London, 1949.
10. "Catalyst Handbook," p. 169, Springer Verlag, New York, 1970.
11. General Motors Corp., unpublished data.
12. Barnes, G. J., Klimisch, R. L., Krieger, B. B., Int. Automot. Eng. Congr., Detroit, January, 1973, paper **730200**.
13. McArthur, D. P., "Activity, Selectivity, and Degradation of Auto Exhaust NO_x Catalysts," N. Amer. Mtg. Catal. Soc., 3rd, San Francisco, February, 1974.

RECEIVED May 28, 1974.

6

Platinum Catalysts for Exhaust Emission Control: The Mechanism of Catalyst Poisoning by Lead and Phosphorus Compounds

G. J. K. ACRES, B. J. COOPER, E. SHUTT, and B. W. MALERBI

Johnson Matthey & Co. Ltd., Research Laboratories, Wembley, England

The effect of lead and phosphorus on platinum catalysts used to meet United States car exhaust emission legislation is assessed. Whilst phosphorus was the most toxic element on a w/w basis, interaction between phosphorus and lead compounds resulted in the trace lead in the fuel contributing the major effect. Evaluation of lead poisoning using tetraethyllead indicates a first order deactivation process with adsorption of lead being controlled predominantly by degradation of tetraethyllead prior to arrival at the catalyst surface. It is suggested that adsorbed lead may consist of particulate and nonparticulate species; the former has relatively little effect on catalyst performance whereas the latter adsorbs on active sites resulting in significant degradation even at levels less than 0.05 g lead/U.S. gal.

The successful use of platinum monolithic oxidation catalysts to control automobile emissions over many thousands of miles requires an intimate understanding of the many factors which contribute to catalyst degradation. Contamination of the active catalyst by lead and phosphorus compounds present in fuel and lubricating oil is a major factor in catalyst deterioration.

Catalyst deactivation by lead and phosphorus compounds in the fuel was reported by Gagliardi *et al.* (*1*), who demonstrated that phosphorus was potentially more hazardous than lead on a w/w basis. Shelef *et al.* (*2*) demonstrated that there is a synergistic effect between lead and phos-

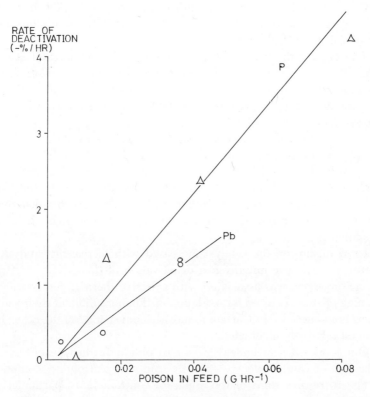

Figure 1. The rate of catalyst deactivation as a function of poison in the feed

Simulated test; ○, Pb as tetraethyllead and △, P as tri-n-butyl o-phosphate

phorus which occurs in the exhaust system and which renders the compounds less toxic, possibly by the formation of lead phosphate.

Partial recovery of automobile catalyst from lead and phosphorus deactivation was reported (1), and chemical cleaning of catalyst surfaces by acid leaching and ultrasonics reactivated partially poisoned catalysts (2).

Commercially available lubricating oils caused relatively little deactivation in comparison with lead and phosphorus contamination from fuel. However, Acres and Cooper (4) demonstrated that when the normal dialkylzinc dithiophosphate additive was replaced by similar compounds without the heavy metal—such as in some forms of ashless oil—catalyst deactivation was rapid. Again the nontoxicity of phosphorus compounds in the presence of heavy metal was associated with formation of inorganic phosphates.

Other materials in fuel and lubricating oil, such as sulfur (*1, 4*) and compounds of the alkaline earths (*4*), caused little poisoning relative to that of lead and phosphorus.

The greater the lead level in the fuel, the faster the catalyst deactivation (*2*). The amount of lead deposited on the catalyst was proportional to that in the fuel and the total lead passing through the engine (*3*). Approximately 15% of the lead burned was retained on the catalyst samples (*2*).

Using pulse flame burners to simulate fuel combustion, McDonnell *et al.* (*3*) established that catalyst activity is a logarithmic function of the catalyst's lead content; correlation of laboratory studies with actual vehicle use was excellent. However, their correlation was limited to fuels with < 0.003–0.05 g Pb/U.S. gallon (USG), and greater deactivation occurred at 0.5 g Pb/USG even though the lead content of the catalyst was similar.

Lead pickup on the catalyst increased with increasing catalyst temperature (*1, 2*). The initially fast deactivation of the catalyst for hydrocarbon conversion was associated with selective thermal degradation and poisoning of sites required for oxidation of the more difficult hydrocarbons such as methane (*2, 3*). Carbon monoxide conversion, on the other hand, remained relatively unaffected.

Our work was an endeavor (a) to obtain further insight into the mechanism of catalyst deactivation by lead and phosphorus compounds, and (b) to resolve some of the apparent contradictions, particularly the relation between lead deposits and catalyst activity.

Experimental

Both simulated exhaust apparatus and single-cylinder engines were used with tetraethyllead (TEL) and tributyl *o*-phosphate (TBP) as the test poisons. No halide scavengers were present in the fuel used for the engines to enable correlation with the findings from the simulated tests.

The simulated test was designed to dope a catalyst sample with a given amount of lead by atomizing the poison compound (TEL) into a hot, simulated exhaust gas stream just prior to the inlet face of the catalyst sample. A liquid hydrocarbon carrier served as a convenient means of introducing TEL into the exhaust stream *via* an atomizer, and it also provided the hydrocarbon constituent of the simulated exhaust gas. *n*-Heptane was chosen as the carrier since it is representative of average fuel hydrocarbon compositions.

The simulated exhaust test apparatus consisted of a tubular reactor fitted with a catalyst chamber 1.9 in. in diameter and 3 in. long. The hydrocarbon feed containing dissolved poison was atomized into a premixed gas which had been preheated by passage over an electric heater. The heater was controlled by a thermocouple placed at the inlet face of the catalyst; it raised the exhaust gas temperature to a preset level to give the desired catalyst inlet temperature. At this stage, oxygen was absent from the exhaust to prevent oxidation of reactants on the heating coils. Oxygen was intimately admixed with the exhaust just prior to the catalyst sample.

A typical gas arriving at the catalyst face contained 2.5% CO, 1000 ppm HC, 500 ppm NO_x, 3% O_2, and 0.48 g Pb/hr (as tetraethyllead). Catalyst inlet and exit gas samples were monitored for HC by flame ionization detector, for CO by nondispersive IR analyzer, and for O_2 by paramagnetic analyzer. The effect of the TEL poison on catalyst activity was monitored by measuring the decrease in CO and HC conversion levels.

The simulated test did not deliberately attempt to precombust the poison compound prior to injection over the catalyst, and thereby differed from the approach of Shelef et al. (2). Therefore, comparison of engine and simulated tests would be of interest since this could reflect differences in the nature of the poisoning species. Furthermore, since the test simulates engine exhaust with n-heptane as the hydrocarbon, methane is absent. In real engine exhaust, methane plays a predominant role in the rate of loss of HC activity during the initial stages of catalyst deactivation (2, 3). Therefore, in order to remove methane conversion activity to enable correlation of engine and simulated test runs, the catalysts were thermally aged at 950–1000°C for 1 hr prior to the start of poisoning. Lead levels in both tests were generally higher than those associated with normal catalyst operation in order to accelerate the poisoning effect.

Engine tests were performed on single-cylinder, 175-cc engines coupled to electrical generators so that a load could be applied to the engine by increasing the electrical load. The catalyst pieces were 1.9 in. in diameter and 3 in. long. An in-line electric heater on the engine exhaust controlled inlet gas temperatures to the catalyst. Composition of the exhaust from the single-cylinder engines was typically 2.5–5.5% CO, 3.0–6.2% O_2, 500–2000 ppm NO_x, 500–2000 ppm HC. Catalyst inlet and exit gas compositions were monitored in the same manner as in the simulated tests.

Results

Comparison of Phosphorus and Lead as Poisons. In order to assess the relative toxicities of the two major poisons of noble metal oxidation catalysts, simulated tests were run with the loss of catalyst activity for HC conversion measured as a function of the TBP and the TEL in the feed, these being representative of additives in commercial fuels.

Table I. Comparison of Oil and Fuel Additives as Catalyst Poisons

In Fuel, g/gal		In Oil, wt %		HC Conversion
Pb	P	Zn	P	at 100 hrs, %
sterile		0.085	0.070	78
sterile		0	0.060	57
0.05	0	0.085	0.070	63
0.05	0.01	0.085	0.070	77

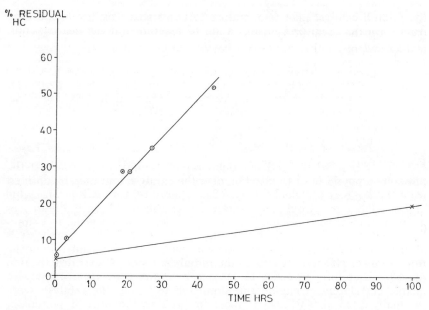

Figure 2. The rate of catalyst deactivation as a function of concentration of lead in the fuel

Single-cylinder engine test; ×, 0.05 g Pb/USG and ○, 3.0 g Pb/USG

The rate of catalyst deactivation was directly proportional to the poison concentration (Figure 1). Phosphorus was more toxic than lead on a w/w basis by a factor of about 1.5; on an atomic basis, lead was more toxic by a factor of about 4.

Poison Interactions. Lead and phosphorus interacted when mixed together in the fuel, and the resultant rate of deactivation was less than that when either was used singly (2). This effect has major implications in the evaluation of exhaust control catalysts, both in the design of oils compatible with catalyst systems and in the lead and phosphorus levels specified in the fuel for certification. Therefore, a series of tests was conducted with single-cylinder engines in order to evaluate these interactions. Data on the hydrocarbon conversion efficiency of the catalyst after 100 hrs running (Table I) clearly demonstrated that deactivation was more serious when lead or phosphorus was present alone, whereas catalyst activity was maintained when both lead and phosphorus were present in the fuel and when phosphorus was combined with zinc in the oil.

The Effect of Lead Content in the Fuel—Static Engine Tests. The effect of increasing lead content in the fuel on HC conversion efficiency in single-cylinder engines is depicted in Figure 2. Loss in activity was dramatic at the higher lead level.

Figure 3 depicts the loss in activity at various lead levels and lead pickup on the catalyst samples. Loss in catalyst activity was approximately proportional to the amount of lead deposited on the catalyst.

The lead-loading effect was also confirmed on a 1.8-liter, 4-cylinder engine (Figure 4); catalyst deactivation was first order with respect to the lead content of the fuel (in the region investigated). It should be noted that control of hydrocarbon emissions by catalytic oxidation was more difficult than control of carbon monoxide emissions.

The Effect of Lead Content in the Fuel—Road Endurance Tests. Emission tests during road mileage accumulation using noble metal, monolithic catalysts demonstrated again that the major deterioration in emission occurs with respect to the hydrocarbon component. Assessment of hydrocarbon emissions during tests using fuels containing 0.01 and 0.05 g Pb/USG (Figure 5) revealed that a deterioration in hydrocarbon emissions occurred with both, but the risk of catalyst deterioration was more severe with the latter. Thus, although catalyst deactivation was observed with the lead sterile fuel (< 0.01 g Pb/USG) as a consequence of numerous factors including thermal deactivation and attack by other additives, it would appear that increasing the lead content to 0.05 g/USG significantly increased the rate of catalyst deactivation. The consequence of this effect is revealed by extrapolating the data to the permitted 1976 U. S. Federal limit of 0.41 g hydrocarbon emissions per mile. Whereas with 0.05 g Pb/USG the limit would be exceeded after only 10,500 miles, with fuel containing less than 0.01 g Pb/USG the limit would not be

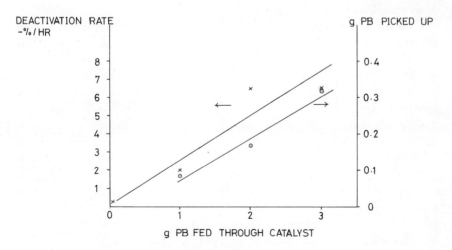

Figure 3. Catalyst deactivation and lead pickup as a function of combusted lead

Single-cylinder engine tests; ×, rate of deactivation and ○, lead pickup

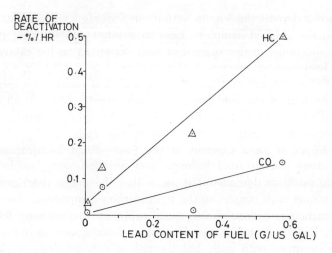

Figure 4. Rate of catalyst deactivation as a function of lead content in the fuel

Four-cylinder, 1.8-liter engine test; △, HC and ○, CO

exceeded until 33,000 miles with the particular engine–emission–catalyst system used on this test vehicle.

Spent Catalyst Analysis. Analysis of catalysts subjected to car and engine testing by x-ray fluorescence (XRF) revealed the expected con-

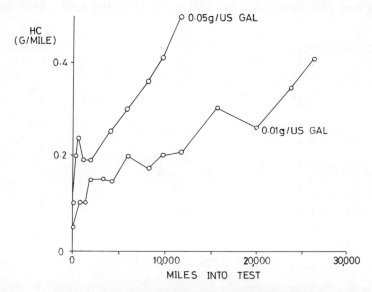

Figure 5. Road endurance trials performed at two levels of lead

Chrysler/Avenger/Plymouth Cricket test vehicle

Table II. Lead and Phosphorus Accumulation on Johnson Matthey Oxidation Catalysts after Engine Testing

Lead Content of Fuel, g/USG	Pb Deposited on Catalyst: Pb Consumed by Engine	Lube Oil Consumed, pints	P Deposited on Catalyst: P Consumed by Engine
0.01	0.26	10.5	0.16
0.05	0.31	12.5	0.17
0.30	0.24	6.0	0.12
0.57	0.25	12.5	0.13

tamination by lead and phosphorus compounds. When the catalyst was subjected to various levels of lead in the fuel feed, the fraction of combusted lead accumulated on the catalyst block remained virtually constant (Table II) as did the phosphorus pickup from the lubricating oil.

There was marked radial and axial distribution of lead and phosphorus within the catalyst bed. The radial poison distribution was almost

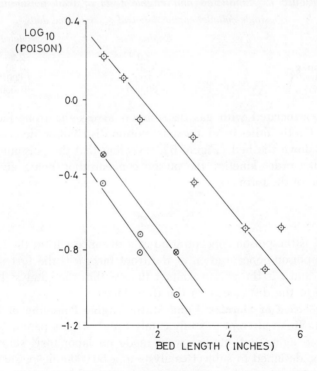

Figure 6. The first order dependence of lead and phosphorus deposition

⊙, phosphorus on the edge of bed (Avenger trial); ⊗, phosphorus on the center of bed (Avenger trial); and ⊖, lead on the center of bed (1.8-liter, static engine test)

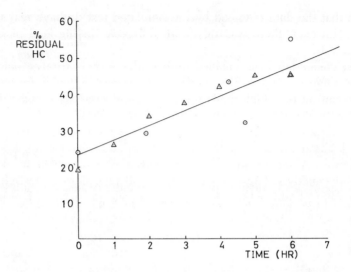

Figure 7. Simulated and engine tests of lead poisoning
○, single-cylinder engine data and △, simulated data

	Engine Test	Simulated Test
Lead fed	3.0	2.9
Lead pickup, g	0.37	0.20
Temperature, °C	500	600
Space velocity, hr^{-1}	80,000	91,000

certainly associated with gas distribution across the front face of the catalyst. On the other hand, the axial poison distribution decreased exponentially down the bed (Figure 6), revealing that the poisoning process follows first order kinetics, the poison concentration being given by an expression of the form:

$$C_x = C_o e^{-kt}$$

where C_x is the poison concentration at a distance x from the front face, C_o is the poison concentration at the front face, k is the first order rate constant, and t is the poison contact time within the catalyst bed (proportional to the distance from the front face).

Correlation of Simulated and Static Engine Poisoning of Platinum Catalysts. In an attempt to study catalyst and process parameters by an accelerated aging test, studies were made on laboratory, simulated exhaust rigs, designed to subject catalysts to a 50,000-mile poisoning durability test in a single day. Obviously, good correlation of simulated and engine performance is essential. The correlation achieved under conditions compatible to both engine and simulated rigs is depicted in Figure 7. Correlation of catalyst performance was good, and this imparted confidence that the data obtained by the simulated test do have real significance. However, there was significant difference in the lead pickup in the two tests.

Use of simulated rigs enabled study of the mode of catalyst poisoning since evaluation of poisoning deactivation curves provides insight into the mechanism by which noble metal oxidation catalysts become deactivated by the acquisition of poisons from the gas phase.

The typical deactivation curves plotted in Figure 8 depict the poisoning of Johnson Matthey platinum metal oxidation catalyst by lead and phosphorus. A section of the spent catalyst was subjected to electron probe microanalysis (Figure 9); the lead and phosphorus density photographs were obtained by monitoring back reflected x-rays. Phosphorus accumulated on the periphery of the washcoat at the gas–solid interface, whilst lead appeared to be more evenly distributed throughout the

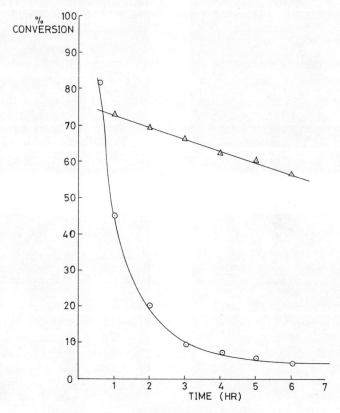

Figure 8. Typical poisoning deactivation curves
Simulated experiments; ○, phosphorus and △, lead

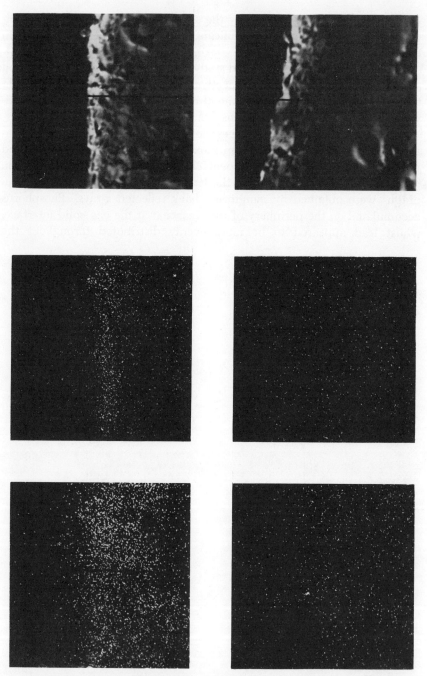

Figure 9. Electron probe microanalysis of spent oxidation catalysts

Magnification, ×576; left, *used catalyst;* right, *unused catalyst;* top, *electron images;* middle, *phosphorus x-ray images;* and bottom, *lead x-ray images*

washcoat phase and there was some evidence that the monolith structure itself had been penetrated.

Lead Poisoning as a Function of Engine Load. A preliminary study of the effect of engine load on catalyst poisoning by lead was made on single-cylinder engines. The rate of poisoning increased with engine load-

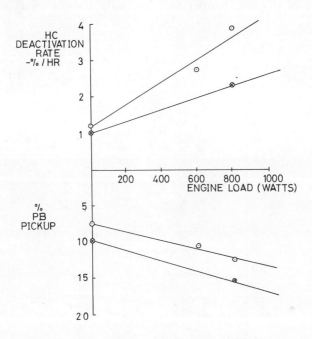

Figure 10. Rate of catalyst deactivation and lead pickup as a function of engine load

Single-cylinder engine; ⊙, catalyst 1 and ⊗, catalyst 2

ing as did the lead pickup (Figure 10). The findings were not quite clear-cut since the measured rates of deactivation lie within experimental scatter, but the indications are that the effect is real and this was confirmed by the lead pickup analyses on the used catalysts.

The Effect of Catalyst Temperature. Simulated rig data on the effect of catalyst temperature are plotted in Figures 11 and 12. The rate of catalyst poisoning was increased by higher catalyst bed temperatures. Care must be exercised in assessing the findings since the oxidation activity of the catalyst increases with increasing operating temperature (kinetic effect), and this must be allowed for in the poisoning measurements. For example, when the poisoning rate was measured at the temperature of the run, changes in the rate of catalyst deactivation between 500° and 700° were significant only toward the end of the 700°C run

(Figure 11). However, when measurements were made at a datum temperature common to each run, the anomaly disappeared (Figure 12).

In single-cylinder engine tests, measurements were made at the temperature of the test. With 1 g Pb/USG (Figure 13), the predominant effect was that of temperature on residual hydrocarbons, and, under the conditions of the test, temperature had only a slight effect on the rate

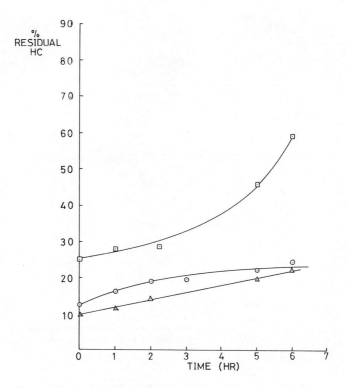

Figure 11. Catalyst poisoning by lead as a function of catalyst bed temperature

Simulated test with measurements made at bed temperature; ○, 500°C; △, 600°C; and ☐, 700°C

of catalyst deactivation by lead poisoning. This illustrates how a kinetic effect compensates for the enhanced rate of catalyst poisoning at high temperatures. However, at 3 g Pb/USG the kinetic effect could not compensate for the enhanced poisoning rate (Figure 14).

The lead pickup data (Figure 15) indicated that with 1 g Pb/USG retention was independent of temperature, and the poisoning rate increased only slightly with temperature (*see* Figure 13). At these lead levels, we are operating in a region of good correlation between catalyst

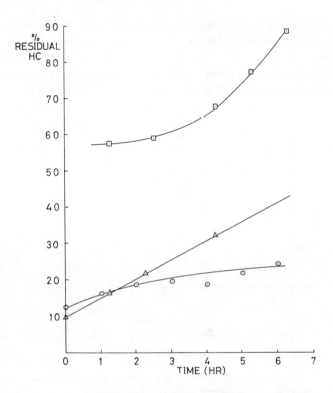

Figure 12. Catalyst poisoning by lead as a function of catalyst bed temperature

Simulated test with readings taken at datum temperature of 500°C; ○, 500°C; △, 600°C; and □, 700°C

deactivation rate and lead pickup (*see* Figure 3). With 3 g Pb/USG in the fuel, lead pickup increased with temperature as did the rate of deactivation (Figure 14).

Discussion

The Relative Poisoning Powers of Lead and Phosphorus Additives. The rate of deactivation of platinum oxidation catalyst leads to a first order dependence on lead concentration at low lead levels (*see* Figure 3).

Acres and Cooper (4) demonstrated that only lead and phosphorus were potential poisons to platinum oxidation catalysts in this application and that phosphorus was more toxic on a weight basis, with lead more toxic on an atomic basis. Furthermore, there is strong evidence that lead and phosphorus interact both in the engine upon combustion and on the catalyst. Thus fuels containing both these elements produce lower rates of deactivation than when each is used singly. It is concluded that the interaction produces lead phosphate particulates which are only loosely held or physically adsorbed by the catalyst, and which may be regarded as nontoxic.

Catalyst deactivation from a poisoning mechanism was most severe when lead alone was present in the fuel. Hence lead in the fuel may be

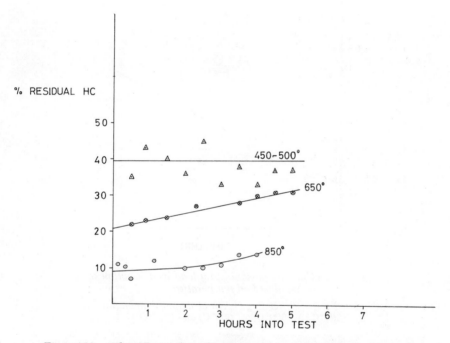

Figure 13. The effect of catalyst bed temperature on HC conversion efficiency using fuel containing 1 g Pb/USG

Single-cylinder engine; △, 450°C; ⊗, 650°C; and ⊙, 850°C

considered the major threat to catalyst durability. Similarly phosphorus in the engine lubricating oil is no problem to platinum oxidation catalysts when it is present as dialkylzinc dithiophosphate. On the other hand, organophosphorus additives (*e.g.*, in ashless oils) may be deleterious to catalyst durability.

Poison Deposition on the Catalyst. XRF analysis of spent catalyst units revealed that only a small percentage of the lead and phosphorus burned in the engine was retained by the catalyst. This agrees with the concept of particulate formation upon combustion in the engine with resultant low retention on the catalyst. Most of the lead and phosphorus species analyzed on the catalyst probably are present as particulates, and toxic species on the catalyst constitute only a very small portion of the

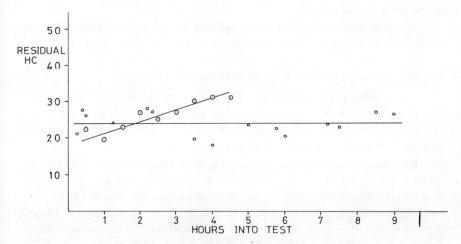

Figure 14. The effect of catalyst bed temperature on HC conversion efficiency using fuel containing 3 g Pb/USG

Single-cylinder engine; ○, 650°C and ⊙, 850°C

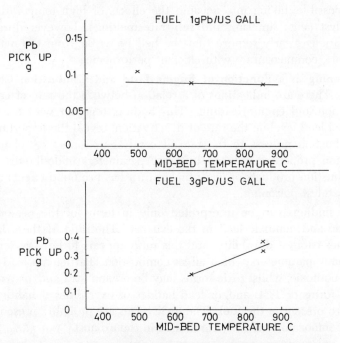

Figure 15. Lead pickup as a function of catalyst bed temperature and concentration of lead in the fuel

lead burned in the engine. However, since oxidation catalysts are degraded by lead, some organo and/or volatile lead compounds are present in the exhaust gas stream.

We also identified a first order relation for poison deposition on the catalyst. Therefore, accumulation of poison on any element of catalyst is directly proportional to the local concentration of poison in the gas phase, and the poisoning process may be diffusionally controlled.

The quantitative assessment of axial poison distribution (when the first order rate constant and percentage poison pickup are known accurately), together with knowledge of radial distributions, will allow optimization of catalyst design for poison resistance.

Poisoning as a Function of Catalyst Bed Temperature. The rate of catalyst poisoning was enhanced by high bed temperatures. It is likely that this is attributable solely to the effect of temperature on the poisoning mechanism since the catalysts were thermally pretreated prior to poisoning. However this does not eliminate the possibility of cooperative thermal interaction between the lead species and the catalyst–support surface causing loss in catalyst activity. These findings have implications in locating oxidation catalysts relative to the exhaust manifold. Cold start considerations demand that the catalyst should be as close to the manifold as possible, taking into account the effects of high temperature on the catalyst (e.g., sintering and substrate melting); however durability considerations clearly require that the bed be as far from the manifold as possible, commensurate with efficient performance.

Poisoning as a Function of Engine Load and Combustion Characteristics. There are indications of a relation between the rate of catalyst deactivation and engine loading. This finding, together with the effect of the fuel lead level on the rate of deactivation, reveal the importance of the combustion process to the rate of catalyst poisoning. A change in combustion properties induced by these parameters undoubtedly alters the combustion products of lead; this in turn has profound effects on the rate of catalyst poisoning.

The findings can be interpreted only in terms of the presence of both toxic and nontoxic lead in the exhaust. The ratio of these largely determines catalyst durability, and this ratio appears to be very sensitive to the many parameters which affect combustion. Particulate lead products are nontoxic, whilst toxic forms may be organo lead and/or partially oxidized forms of TEL and/or lead halides or oxyhalides if halide scavengers are present in the fuel. The toxic forms are probably present only as a very minor portion of the total lead in the exhaust.

The Mechanisms of Poisoning. The curves in Figure 8 depict typical deactivation characteristics of oxidation catalysts poisoned by lead and phosphorus separately in sir lated exhaust tests. Such curves may be interpreted in terms of poisoning in porous catalyst structures by the Wheeler method (5).

The phosphorus deactivation curve is typical type C, and, according to the Wheeler model, this is associated with selective poisoning of pore mouths. Phosphorus distribution on the poisoned catalyst is near the gas–solid interface, *i.e.* at pore mouths, which confirms the Wheeler model of pore mouth poisoning for type C deactivation curves. Thus we may propose that in the fast oxidative reactions with which we are dealing, transport processes within pores will control the effectiveness of the catalyst. Active sites at the gas–solid interface will be controlled by relatively fast bulk diffusional processes, whereas active sites within pores of 20–100 A present in the washcoat aluminas on which the platinum is deposited will be controlled by the slower Knudsen diffusion process. Thus phosphorus poisoning of active sites at pore mouths will result in a serious loss in catalyst activity since reactant molecules must diffuse deeper into the pore structure by the slower Knudsen mass transport process to find progressively fewer active sites.

On the other hand, lead poisoning deactivation curves have nonselective characteristics (Type A). These result when the poison is less strongly chemisorbed, and it tends to suffer many collisions with the alumina washcoat structure before chemisorption. Consequently, lead is found deep inside the washcoat structure, as is demonstrated by electron probe microanalysis, and the more accessible metal sites are left active to gaseous reactants by the faster bulk transport processes.

Literature Cited

1. Gagliardi, J. C., Smith, C. S., Weaver, E. E., Amer. Petrol. Inst. Div. Refining, Mid-year Mtg., 37th, New York, May, 1972, paper **63–72**.
2. Shelef, M., Dalla Betta, R. A., Larson, J. A., Otto, K., Yao, H. C., "Poisoning of Monolithic Noble Metal Oxidation Catalysts in Automobile Exhaust Environment," Amer. Inst. Chem. Eng., Nat. Mtg., 24th, New Orleans, March, 1973.
3. McDonnell, T. F., McConnell, R. J., Otto, K., Amer. Petrol. Inst., Div. Refining, Mid-year Mtg., 38th, Philadelphia, May, 1973, paper **14–73**.
4. Acres, G. J. K., Cooper, B. J., Proc. Int. Clean Air Congr., 3rd, Dusseldorf, October, 1973, **F14**.
5. Wheeler, A., Advan. Catal. (1951) **3**, 309.

RECEIVED May 28, 1974.

7

A Comparison of Platinum and Base Metal Oxidation Catalysts

GERALD J. BARNES

Physical Chemistry Dept., Research Laboratories, General Motors Corp., Warren, Mich. 48090

The activities of fresh, supported platinum and base metal oxidation catalysts are evaluated in vehicle tests. Two catalysts of each type were tested by the 1975 FTP in four 600–4300 cm^3 catalytic converters installed on a vehicle equipped with exhaust manifold air injection. As converter size decreased, base metal conversions of HC and CO decreased monotonically. In contrast, the platinum catalysts maintained very high 1975 FTP CO conversions ($>90\%$) at all converter sizes; HC conversions remained constant ($\sim 70\%$) at volumes down to 1300 cm^3. Performance of the base metal catalysts with the 4300-cm^3 converter nearly equalled that of the platinum catalysts. However, platinum catalysts have a reserve activity with very high conversions attained at the smallest converter volumes, which makes them more tolerant of thermal and contaminant degradation.

The use of supported, platinum group, metal oxidation catalysts to control automotive exhaust hydrocarbon (HC) and carbon monoxide (CO) emissions is expected to commence with 1975 model year vehicles. Because of the cost of these metals and questions about their availability, base metal catalysts were considered an attractive alternative.

These experiments provide a direct comparison of the initial activities of platinum and base metal catalysts. Differences in performance— produced by such variables as catalyst bed mass, exhaust gas space velocity, and catalyst temperature—are explained by the effect of converter size on warm-up rates and by the kinetic differences for oxidation reactions over the two types of catalysts.

Table I. Baseline Values for Vehicle Emissions[a]

1975 FTP Emissions		Cycle 1-5 Emissions		Cycle 6-18 Emissions	
HC, g/mile	CO, g/mile	HC, g	CO, g	HC, g	CO, g
0.98	36.5	4.34	200.8	3.77	127.0
1.10	36.8	5.47	204.0	4.03	131.4
0.93	39.0	5.70	210.4	3.44	144.4
0.99	39.6	3.61	179.0	3.60	137.5
1.01	35.7	5.18	173.0	3.07	130.2
0.91	37.9	4.91	187.4	3.71	134.0
0.86	43.6	5.02	174.5	2.73	156.3
0.95	40.6	5.37	180.6	3.18	137.0
0.84	41.5	4.44	176.1	3.02	137.6
Average 0.95	39.0	4.89	187.3	3.39	137.3

[a] Measured with the air injection system in operation, but without a catalyst.

Experimental

Test Vehicle. The test vehicle (1) was equipped with a 5.7-liter displacement, V8 engine with carburetion adjusted for slightly richer than stoichiometric air/fuel ratios. This modification necessitated the use of an air pump to provide sufficient oxygen (O_2) for control of exhaust HC and CO. The air pump output was injected into the exhaust manifolds ahead of the catalytic converter. This point of air injection provided sufficient time for mixing of the air and exhaust gases before they entered the catalytic converter, and it also caused appreciable homogeneous oxidation of HC and CO in the exhaust manifolds.

Base line values for vehicle emissions (with air injection but without catalyst) were measured periodically during the catalyst test series. Base line values were determined with empty converters as well as with inert SiC filler pellets, and there was no discernible difference. The repeatability of the base line values is evident in Table I. Base line emissions are given for the 1975 Federal Test Procedure (FTP), for the first five cycles of operation (Bag 1), and for the hot, stabilized cycles 6–18 (Bag 2). These values thus reflect the overall vehicle emissions, emissions during warm-up, and emissions during stabilized vehicle operation, respectively. Catalyst system performance was assessed by comparison with the average base line emission values, *i.e.* the latter were considered the converter inlet levels of HC and CO.

The fuel used in all the tests was clear (unleaded) Indolene which contained about 0.05 g lead/gal, 0.003 g phosphorus/gal, and 0.02 wt % sulfur.

Emission Measurement. Exhaust gases leaving the vehicle tailpipe were diluted and cooled by addition to a stream of air in the constant volume sampler. Exhaust emission concentrations in the diluted stream were monitored continuously as the vehicle was operated over the 1975 FTP (*see* Table II for instrumentation). Instrument output was scanned every 0.5 sec, and exhaust mass emissions corrected for temperature and pressure were calculated by numerical integration by digital computer (2).

Table II. Exhaust Emission Monitoring

Exhaust Constituent	Instrument
CO	nondispersive IR analyzer
CO_2	nondispersive IR analyzer
HC	flame ionization detector
NO_x	chemiluminescent analyzer

Except for preconditioning the evaporative emission control canister, the emission tests were performed as prescribed (3, 4, 5, 6). As an expedient, test variability attributable to variable HC loading on the evaporative control canister was eliminated by using a purged canister for all tests. All cold start tests were performed after a minimum 12-hr vehicle soak time.

Table III. Catalytic Converters

Volume, cm^3	Bed Depth, cm	Frontal Area, cm^2	Bed Configuration	Catalyst Mass,[a] g
600	5	120	skewed rectangular	330
1300	5	260	skewed rectangular	715
3000	5	600	skewed rectangular	1650
4300	5	860	rectangular, rounded ends	2365

[a] Based on a typical, packed density of 0.55 g/cm³.

Catalytic Converters. The catalytic converters were of experimental design, with volumes of 600–4300 cm³ (Table III). The three smaller converters were simple stainless steel shells, with the pelleted catalysts retained by 10-mesh stainless steel screens. The largest converter was a production prototype of somewhat more rugged and hence more massive design. The converters were installed just downstream of the exhaust system Y where exhaust from the two banks of the engine merges. They were therefore under the floor of the right front passenger compartment of the vehicle.

Catalyst Properties. Of the four catalysts evaluated, two were prepared at the General Motors Research Laboratories (GMR) and two were commercially formulated, with one base metal and one platinum catalyst in each group (Table IV). All the catalysts were bulk, pelleted types

Table IV. Catalysts

Catalyst	Active Component, wt %
GMR Pt	0.3 Pt
PT	Pt[a]
GMR Cu–Cr	8.0 Cu + 7.0 Cr
BM	base metal[a, b]

[a] Unspecified concentration.
[b] Unspecified composition.

with spherical supports of nominal 3-mm diameter. The GMR catalysts used preformed alumina pellets (Kaiser KC/SAF, gel derived); the commercial supports were undisclosed for proprietary reasons. Similarly, the nature and the amount of active component in the two commercial catalysts (PT and BM) are not known. The GMR catalysts contained 0.3 wt % Pt (GMR Pt) or a mixture of Cu and Cr oxides with 8 wt % Cu and 7 wt % Cr (GMR Cu–Cr).

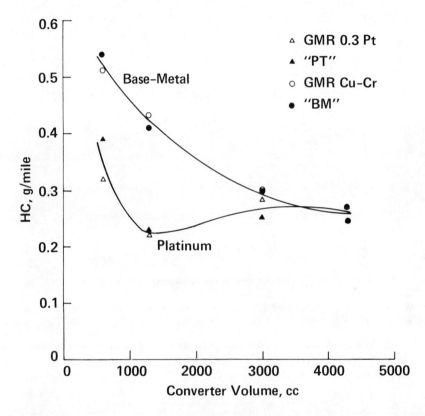

Figure 1. 1975 FTP HC emissions

Results

Sixteen combinations of converter size and catalyst composition (four catalysts in each of four converter volumes) were evaluated. A fresh charge of catalyst was placed in each converter, and the converter was then installed on the vehicle. Three repeat 1975 FTP emission tests were run with no other mileage accumulation on the catalyst. Test results were then averaged, the average representing the zero-miles activity of the catalyst for that converter size.

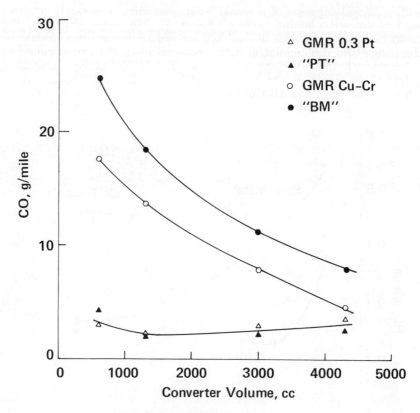

Figure 2. 1975 FTP CO emissions

1975 FTP Emissions. The overall system performance during the 1975 FTP tests for HC and CO emissions as a function of catalytic converter volume is plotted in Figures 1 and 2 respectively. Emissions are expressed in g/mile of vehicle operation with the cold start, hot stabilized, and hot cycle emissions weighted as prescribed (3). There was a distinct difference in the performance of the base metal and platinum catalysts with decreasing converter volume. Both HC and CO emissions with base metal catalysts increased monotonically as converter volume was decreased. In contrast, when platinum catalysts were used, both HC and CO emissions decreased to a minimum at ~1300 cm³ and then increased at the smallest volume.

Warm-Up Performance. 1975 FTP emissions are strongly affected by vehicle emissions during warm-up, particularly when catalyst systems are used to control emissions. The effect of catalyst volume, and thus the thermal mass of the converter, on warm-up temperatures is depicted in Figures 3 and 4 for the GMR Pt and Cu–Cr catalysts respectively.

These temperatures, the average from two thermocouples placed in the midplane of the converter, are plotted as a function of time. The time span included the first two cycles of the FTP.

For both catalysts, the rate of increase in bed temperature was greater for smaller converter volumes although there was little difference between the 600- and 1300-cm^3 converters. Catalyst temperature increased more rapidly with the platinum catalyst. Greater conversion efficiencies (Table V) and hence greater energy release rates were attained with this catalyst during warm-up cycles 1–5 of the FTP. The fraction converted by the platinum catalysts was maximum at the 1300-cm^3 converter volume. Conversion efficiencies with the base metal catalysts increased continuously as volume increased.

Hot Cycle Emissions. The 1975 FTP results were strongly affected by the warm-up performance of the system. The hot cycle emissions, after catalyst temperatures stabilized, are more directly related to the effect of converter volume on performance. Hot cycle (FTP cycles 6–18) HC and CO emissions are depicted in Figures 5 and 6 respectively.

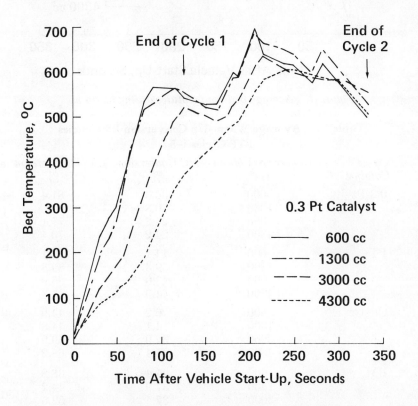

Figure 3. Average bed temperatures during warm-up

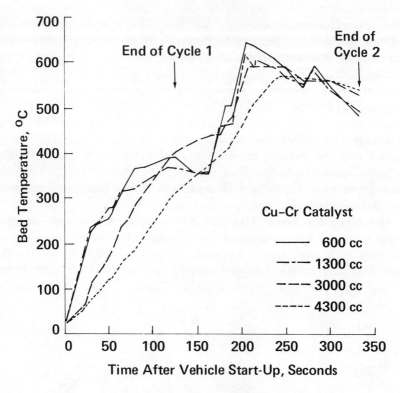

Figure 4. Average bed temperatures during warm-up

Table V. Average Warm-Up Conversion Efficiencies FTP Cycles 1–5

Catalyst	Converter Volume, cm^3	HC Conversion, %	CO Conversion, %
0.3 Pt	600	36.3	81.4
	1300	40.8	84.6
	3000	24.2	77.1
	4300	24.2	70.4
PT	600	14.1	78.1
	1300	59.3	89.8
	3000	43.0	85.9
	4300	44.5	81.3
Cu–Cr	600	6.2	44.9
	1300	4.1	44.8
	3000	32.9	59.9
	4300	45.4	68.2
BM	600	24.4	38.8
	1300	37.9	44.2
	3000	37.4	52.5
	4300	47.0	54.9

Conversion efficiencies relative to the base line vehicle emissions are tabulated in Table VI. In general, the response of both types of catalyst to decreasing converter volume was similar to that for the 1975 FTP emissions. Both HC and CO emissions continually increased as base metal catalyst volumes were decreased. HC emissions were relatively constant for platinum catalyst volumes greater than ~1300 cm^3. CO conversions of 97–98% were produced by all combinations of platinum catalyst and converter volume.

In Table VII are listed the average catalyst bed temperatures during two of the hot cycles of the FTP. Cycle 11 is a long (96 sec), relatively constant speed portion of the FTP with mild accelerations and decelerations. A moderate acceleration, a brief cruise, and a normal deceleration constitute cycle 16. Temperatures measured during these cycles are thus representative of hot steady state and transient operations respectively. The very high conversions attained with the platinum catalysts are reflected by these temperatures, which were maximum at a converter

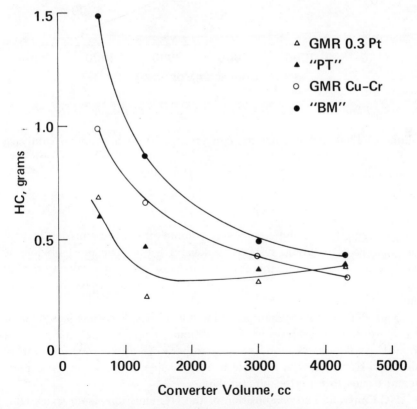

Figure 5. Hot cycle HC emissions during FTP cycles 6–18

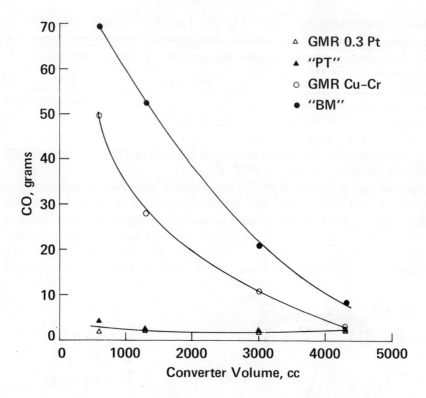

Figure 6. Hot cycle CO emissions during FTP cycles 6–18

volume of 1300 cm³ as were the conversions. With base metal catalysts, temperatures increased with increasing bed volume (and increasing conversion efficiencies) for the largest converter volume.

Discussion

Changing the catalytic converter volume affects the emission control performance of catalytic systems in several ways. The most obvious is that a decrease in volume increases the exhaust gas space velocity during any specific portion of the FTP, with a resultant decrease in reaction time.

The FTP was not designed for kinetic studies, and thus it is difficult to perform kinetic analyses in the traditional manner. Inlet composition, exhaust gas flow rates, and catalyst temperature all vary widely during the FTP. However, it is possible to relate some of the observed experimental trends to reported kinetic data.

HC Control. The composition of the HC emissions from spark ignition engines varies widely with operating mode (7). Over base metal

Table VI. Average Hot Cycle Conversion Efficiencies
FTP Cycles 6–18

Catalyst	Converter Volume, cm^3	HC Conversion, %	CO Conversion, %
0.3 Pt	600	80.0	98.4
	1300	92.6	98.5
	3000	90.9	98.8
	4300	89.1	98.4
PT	600	82.3	96.7
	1300	86.1	98.5
	3000	89.4	98.3
	4300	88.8	98.3
Cu–Cr	600	70.8	64.0
	1300	80.5	79.6
	3000	87.3	92.1
	4300	90.6	97.6
BM	600	56.0	49.9
	1300	74.6	62.1
	3000	85.8	84.9
	4300	87.3	93.7

catalysts, HC oxidation appeared to be first order (8, 9) with the rate constants increasing with carbon number. Olefins and acetylenes had similar rate constants, and paraffins were the most difficult to oxidize catalytically (8). The hot cycle HC conversions over base metal catalysts (Table VI) decreased (and thus emissions increased) as reaction time decreased.

The conversion of exhaust HC over platinum catalysts increased slowly (if at all) with decreasing converter volume, and then decreased rapidly at the smallest volume. The sharp increase in HC emissions at

Table VII. Catalyst Bed Temperatures

Converter Volume, cm^3	Average Temperature, °C	
	GMR Pt	GMR Cu–Cr
Cycle 11		
600	509	366
1300	585	466
3000	498	481
4300	399	389
Cycle 16		
600	512	366
1300	537	457
3000	487	467
4300	387	375

the smallest converter volume might be attributed to the relative ease of catalytically oxidizing non-methane hydrocarbons as opposed to methane. Methane accounts for only 10–25% of the HC exhausted by the engine (7, 10), but 65–75% of the total HC remaining after the catalyst is methane (10). The HC emissions plotted in Figure 5 indicate that nearly complete oxidation of the non-methane hydrocarbons can be achieved with platinum catalyst at volumes greater than 1300 cm^3. The rapid decrease in HC emissions which occurred when the platinum catalyst volume was increased from 600 to 1300 cm^3 would thus be the result of increased oxidation of the non-methane hydrocarbons. Further increases in catalyst volume had little effect on the HC emissions because of the difficulty in oxidizing the remaining methane and because of the small absolute amount of methane involved. This methane limit effect would not be as pronounced over base metal catalysts since the HC conversions did not reach the high levels achieved over platinum catalysts, except at the largest converter volume.

CO Control. Over base metal catalysts, the catalytic oxidation of CO is of the order of 0.7–1.0 in CO with little dependence on O_2 concentration (8, 9, 11, 12). CO oxidation over platinum catalysts is of negative first order dependency on CO concentration (13) with some tendency toward a positive dependency at very high conversion (9). However, this last observation may have been the result of mass transfer limited kinetics (9).

With hot cycle CO emissions, conversion is essentially 100% after the catalyst has warmed up, even at the smallest catalyst volume. Most probably, the small CO emissions during the hot cycles of the FTP are not affected by kinetic effects, but they are related to the transient nature of the vehicle operation during the FTP. An O_2 deficiency could occur if the exhaust gas flows and air pump output were not properly matched. Small CO spikes were observed during vehicle accelerations that were superimposed on a very low, nearly zero, background CO emission level during the hot cycles. There also might be localized O_2 deficiencies if the exhaust gases and air were not well mixed before they entered the converter.

Thermal Effects. As expected, the smaller converter volumes warmed up more rapidly because of their smaller thermal mass (Figures 3 and 4). With the platinum catalysts, the conversion efficiencies achieved during the first five cycles of vehicle operation agreed generally with the 1975 FTP emission trends. A minimum in the emissions (Figures 1 and 2) and a corresponding maximum in the warm-up conversions (Table V) occurred at a converter volume of ~1300 cm^3.

As a result of the very high conversion efficiencies achieved by the platinum catalysts during later portions of the FTP, the hot cycles contributed a negligible mass to the weighted sum of mass emissions which comprises the FTP values. Thus, differences in the FTP findings were determined largely by the warm-up performance of these catalysts.

In later portions of the FTP, the bed temperatures were in good agreement with the trends in the rate of energy release (as indicated by the CO conversion level), except for the largest converter.

Base Metal vs. Platinum Catalysts. When the emission control performances of these two types of catalyst were compared, two points were evident. First, the base metal and platinum catalysts controlled exhaust HC and CO equally (or nearly so) at very large catalyst volumes (low space velocities). This had been reported for laboratory evaluations of these catalysts (14); thus these data for vehicles confirm the findings from steady-state bench tests. Second, initial activity of the platinum catalysts was very high. There was essentially no change in the emissions control performance of the platinum catalysts even at very small catalyst volumes, and this behavior can be utilized in two ways.

First, this reserve activity could be utilized to offset thermal and contaminant degradation of catalyst activity. In simplistic terms, the 4300-cm^3 production prototype converter could lose 70% of its catalytic activity (with a remaining catalyst volume equivalent to 1300 cm^3) without suffering degradation in the overall control of exhaust HC and CO emissions.

Alternately, if thermal and contaminant effects could be minimized, smaller catalyst volumes—and therefore smaller quantities of noble metals—might provide sufficient, long term activity for HC–CO control. Certainly the initial activity of smaller catalyst volumes is sufficient, as these data indicate. This should provide strong incentive for developing more thermally stable catalysts and for minimizing contaminants, most notably those in the fuel supply.

Summary

Supported platinum and base metal catalysts were evaluated in vehicle tests with converter volumes of 600–4300 cm^3. The initial oxidation activity of the catalysts was determined as the vehicle was operated over the 1975 FTP. The ability of the base metal catalysts to control exhaust HC and CO emissions was strongly dependent on the catalyst volume. HC and CO conversion decreased quite rapidly as the converter size was decreased.

The performance of the platinum catalysts was distinctly different. CO conversions remained quite high, even at the smallest catalyst volume. HC emissions increased at the smallest converter size, perhaps because of methane oxidation limitations. CO conversions may have been limited by O_2 availability, as indicated by CO spikes during vehicle accelerations.

Performances of the base metal and platinum catalysts were nearly equivalent at large catalyst volumes. The very high inherent initial oxiddation activity of platinum catalysts indicates that very small catalyst volumes are sufficient for HC–CO control. Larger volumes may even be detrimental to the initial, overall system performance by causing lower conversions during warm-up because of their higher thermal mass. The very high initial platinum activity may also be interpreted as an activity reserve that might be utilized to offset thermal and contaminant degradation effects. These durability effects will probably ultimately determine the required catalyst volume.

Acknowledgments

The author gratefully acknowledges the assistance of J. C. Collins, G. J. Morris, R. W. Richmond, R. M. Sinkevitch, and D. J. Upton in the experimental work. In addition, the discussions on hydrocarbon oxidation with J. C. Schlatter were most helpful.

Literature Cited

1. Barnes, G. J., Klimisch, R. L., "Initial Oxidation Activity of Noble Metal Automotive Exhaust Catalysts," Soc. Automot. Eng., Auto. Eng. Mtg., Detroit, May, 1973, paper **730570**.
2. Mick, S. H., Clark, Jr., J. B., "Weighing Automotive Exhaust Emissions," Soc. Automot. Eng., Mid-year Mtg., Chicago, May, 1969, paper **690523**.
3. *Fed. Regist.* (November 10, 1970) **35** (219), part II, p. 17288.
4. *Fed. Regist.* (March 20, 1971) **36** (55), part II, p. 5342.
5. *Fed. Regist.* (July 2, 1971) **36** (128), part II, p. 12652.
6. *Fed. Regist.* (January 15, 1972) **37** (10), part II, p. 699.
7. McEwen, D. J., *Analyt. Chem.* (1966) **38**, 1047.
8. Innes, W. B., Duffy, R., *J. Air Pollut. Control Ass.* (1961) **11**, 368.
9. Klimisch, R. L., Schlatter, J. C., "The Control of Automotive Emissions by Catalysis," Amer. Ceramic Soc., Flint, September, 1972; General Motors Research Publ. **GMR-1268**.
10. Neal, A. H., Wigg, E. E., Holt, E. L., "Fuel Effects on Oxidation Catalysts and Catalyst-Equipped Vehicles," Soc. Automot. Eng., Auto. Eng. Mtg., Detroit, May, 1973, paper **730593**.
11. Sourirajan, S., Accomazzo, M. A., *Can. J. Chem.* (1960) **38**, 1990.
12. Kuo, J. C., Morgan, C. R., Lassen, H. G., Soc. Automot. Eng., Automot. Eng. Congr., Detroit, January, 1971, paper **710289**.
13. Bond, G. C., "Catalysis by Metals," Academic, London, 1962.
14. Schlatter, J. C., Klimisch, R. L., Taylor, K. C., *Science* (1973) **179**, 798.

RECEIVED June 13, 1974.

8

Deposition and Distribution of Lead, Phosphorus, Calcium, Zinc, and Sulfur Poisons on Automobile Exhaust NO_x Catalysts

DENNIS P. McARTHUR

Union Oil Co. of California, Research Center, Brea, Calif.

> *Catalyst degradation by poisons originating in engine oil and fuel is discussed. Electron probe line scans and x-ray images characterize the variation in contaminant concentrations in both particulate and monolithic NO_x catalysts. Surface contamination was characterized by ESCA. Axial and radial distributions of contaminants were determined for monolithic NO_x catalysts subjected to extensive engine testing. The major contaminants were calcium, lead, phosphorus, and zinc. The concentrations of all contaminants, except possibly lead, varied insignificantly in the radial direction. For all contaminants, the concentration varied greatly in the axial direction, being greatest at the upstream end of the monolith. The degree of contaminant retention on monolithic NO_x catalysts was $P > Pb > Zn > Ca >> S$.*

This is a study of the chemical and physical degradation of NO_x catalysts by contamination. The most detrimental contaminants are those originating from fuel and engine oil, *viz.* lead, phosphorus, sulfur, zinc, calcium, magnesium, and barium. However, engine and exhaust system construction materials can also be harmful catalyst contaminants, *i.e.* iron, copper, nickel, and chrome. NO_x catalysts are normally operated under net reducing conditions whereby the active metals are in a reduced state. Metals in their reduced states are much more reactive than their respective oxides, and they are much more effective "getters" for contaminants. Consequently, it is expected that contaminant poisoning constitutes a more severe problem for NO_x catalysts than for HC–CO oxidation catalysts.

Experimental

The spatial distributions of catalytic metals and contaminant poisons in auto exhaust catalysts were delineated by electron probe line scans. Element concentrations were characterized by element sensitivities, *i.e.* in counts per second (cps). The electron probe microanalyses (EPM) were qualitative or semiqualitative in nature. Accurate correlation between element sensitivity and element concentration requires rather sophisticated instrument calibration. A quantitative evaluation of the EPM findings is beyond the scope of this paper. In general, it can be stated that element concentration is directly proportional to element sensitivity. Furthermore, the proportionality constant between element concentration and element sensitivity varies greatly from element to element.

Sulfur Poisoning

Consider the thermodynamics of metal sulfiding on the basis of the $[H_2S/H_2]$ ratios calculated for synthetic auto exhaust gas (Table I) for an equilibrium reaction system involving the water–gas shift reaction, reactions between H_2, CO, and oxygen, the conversion of SO_2 to H_2S, and metal sulfiding. This reaction system model is as follows:

$$H_2 + 1/2 O_2 \rightarrow H_2O$$
$$CO + H_2O \rightarrow CO_2 + H_2$$
$$C_3H_6 + 3O_2 \rightarrow 3CO + 3H_2O$$
$$SO_2 + 3H_2 \rightarrow H_2S + 2H_2O$$
$$x Me + y H_2S \rightarrow Me_xS_y + y H_2$$

This model assumes that SO_2 in the exhaust gas is completely and very rapidly converted to H_2S.

Table I. Composition of Synthetic Exhaust Gas Used with Bench Scale Flow Reactor[a]

Component	Content, mole %	
	Range	Standard Test
NO	0.01–0.30	0.08
CO	1.0–3.0	2.0
HC	(C_3H_8, C_3H_6)	0.1
CO_2		12.0
H_2O	0–10.0	10.0
O_2	0–1.4	0.35
H_2		0.33
SO_2	0–0.0045	0.0045
N_2		balance

[a] A/F ratio was 12.5–14.6 (value in standard test, 13.8).

Table II. Sulfiding of Metals in Automobile Exhaust
Gas Atmosphere, $[SO_2/H_2]^a \approx [H_2S/H_2]$

Metal	Upper Temperature Limit, °C
Pd, Pt, Ir (IrS$_2$)	232–315
Ru (RuS$_2$)	649–482
Fe	871–621
Co	482–371
Ni (Ni$_3$S$_2$)	899
Cu (Cu$_2$S)	954–593
Pb^b	621–482
Zn^b	always sulfided

^{a} [SO$_2$] = 25 vppm.
^{b} Contaminant poison.

The upper temperature limits for bulk sulfiding of various metals in this reaction system model are listed in Table II. The upper temperature limit is defined as the temperature above which metal sulfiding does not occur. Unless otherwise noted, the form of metal sulfide incorporated into the model was MeS, *i.e.* the monosulfide. On this basis, with a catalyst operating at 538°–815°C, bulk sulfiding is likely to occur with copper, iron, zinc, lead, and ruthenium. However, it has been our experience that used base metal catalysts usually contain less sulfur than a few tenths of one wt %; therefore, bulk sulfiding does not occur.

A chemisorbed monolayer of sulfur on a base metal catalyst (20 wt % base metal) with 50 m^2 of active metal surface area per gram of active metal (~150 Å crystallite size) would be equivalent to ~0.1 and ~0.5 wt % sulfur for monolithic (20 wt % washcoat) and particulate base metal catalysts respectively. For noble metal monolithic catalysts (0.3 wt % noble metal) with 150 m^2 of active metal surface area per gram of active metal (~20 Å crystallite size), an adsorbed sulfur monolayer would be equivalent to ~0.03 wt %. These calculations are based on monolayer sulfur chemisorption, *i.e.* assuming the formation of a surface monosulfide phase. If sulfur were present as a surface oxysulfide or sulfate, the expected catalyst sulfur contents would be significantly lower than the values calculated for monosulfide formation. Thus, the catalyst sulfur contents expected from monolayer sulfur chemisorption (without bulk sulfiding) agree well with the sulfur contents observed in used catalysts.

Sulfur can also be retained on catalysts in the form of sulfates. Many sulfates have good to excellent thermal stability, but it is expected that their hydrothermal stability will vary considerably. Sulfates with contaminant cations which would probably be stable in an auto exhaust gas atmosphere include NiSO$_4$, BaSO$_4$, CaSO$_4$, PbSO$_4$, and nPbO · PbSO$_4$

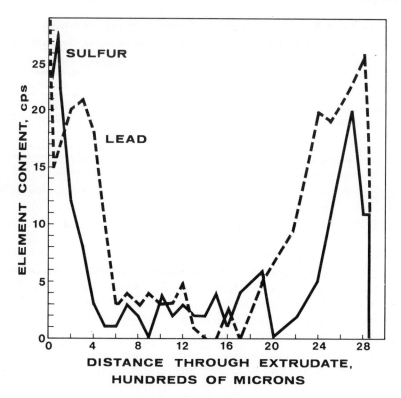

Figure 1. Distributions of lead and sulfur in ⅛-in. extrudate NO_x catalyst after 300-hr engine test

Fuel, 0.06 g Pb/gal and 350 ppm S

(basic lead sulfates). Laboratory experimentation demonstrated that $PbSO_4$ is definitely stable in the exhaust gas environment. Furthermore, routine x-ray diffraction analysis of used auto exhaust catalysts often reveals crystalline $PbSO_4$.

EPM line scans for lead and sulfur were made across the diameter of a ⅛-in. extrudate noble metal catalyst that had been subjected to 300 hrs of engine dynamometer testing using ashless oil and fuel that contained 0.06 g Pb/gal and 350 ppm sulfur (Figure 1). The parallel distributions of lead and sulfur, with concentrations highest at the periphery and then decreasing toward the interior of the catalyst extrudate, suggest the presence of $PbSO_4$. However, $PbSO_4$ could not be detected by x-ray diffraction analysis. Furthermore, similar sulfur distribution curves were also observed when lead was not present on the catalyst.

The distribution of sulfur in a copper–nickel extrudate (1/16-in. diameter) catalyst exposed to synthetic exhaust gas containing 45 vppm SO_2 under varying conditions of time and temperature was detected in

EPM sulfur line scans (Figure 2). The exposure time at each temperature was sufficient to ensure sulfur equilibration between the catalyst and the gas phase with respect to catalyst activity. The catalyst samples were sulfided at temperatures of 427°, 538°, and 677°C (800°, 1000°, and 1250°F), and analysis revealed that they contained 0.32, 0.21, and 0.12 wt % sulfur respectively. In all three samples, the sulfur distribution was symmetrical about a center axis perpendicular to the extrudate cross section. In the samples sulfided at 677° and 538°C, the sulfur concentration was highest at the periphery, and it decreased rapidly toward the center of the extrudate; the depth of sulfur penetration was ~200 and 300μ respectively. The sulfur distribution in the sample sulfided at 427°C was quite interesting. The sulfur concentration was high at the periphery and decreased rapidly to a minimum value at a depth of ~100μ; it then increased to a value approximately equal to that at the periphery and remained fairly constant all the way to the center of the extrudate. Catalyst samples sulfided at 538° and 677°C were brownish and charcoal gray on the outside (skin), and both were black in the interior. However, the catalyst sample sulfided at 427°C had a yellowish-bronze rind (jacket) which extended to a depth of ~100μ; the interior was black. It is particularly interesting that the position of the interface between the yellowish-bronze rind and the black interior coincides with the point of minimum sulfur concentration, *i.e.* at ~100μ inside the periphery. Incidentally, Ni_3S_2 is yellowish-bronze and $NiSO_4$ is yellow.

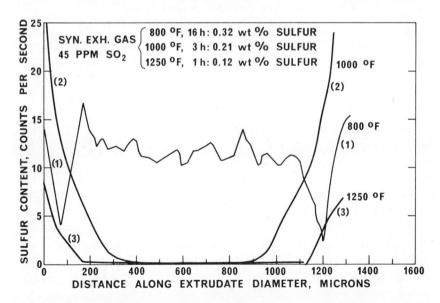

Figure 2. Distributions of sulfur in copper–nickel extrudate NO_x catalyst

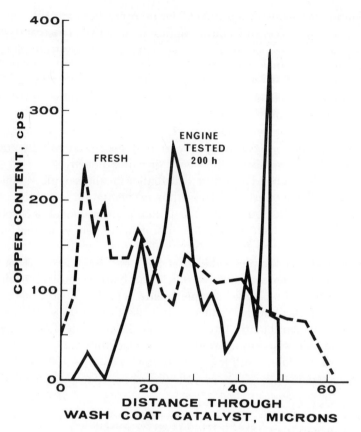

Figure 3. Distribution of copper in copper-promoted monolithic NO_x catalyst before and after 200-hr engine test

Fuel, 0.06 g Pb/gal and 350 ppm S

Metal Mobility

NO_x catalysts are operated at high temperatures under net (often strongly) reducing conditions. The enhanced surface mobility of metals under these conditions must also be considered a cause of catalyst degradation. For example, Figure 3 depicts the distribution of copper in the washcoat of a monolithic catalyst when fresh and after 200 hrs of engine dynamometer testing. This proprietary catalyst consisted of three catalytic metals, one of which was copper, supported on a washcoated cordierite substrate. The washcoat comprised ~15% of the total catalyst weight. Catalyst operating temperature was 649°–704°C.

The copper was fairly uniformly distributed in the fresh catalyst. In the engine-tested sample, the copper was noticeably depleted from the external surface region and concentrated in two bands, one near the middle of the washcoat, the other at the interface between the washcoat and the cordierite substrate. The causes of the observed change in copper distribution are not known at this time. This finding illustrates metal mobility on catalysts operated in an auto exhaust gas atmosphere. The same mobility can be expected for certain contaminant metals, *e.g.* copper and nickel (from construction materials) and lead and zinc (from fuel and engine oil).

The mobility of contaminant metals on NO_x catalysts, especially metals with low melting points such as lead and zinc, is apparent in Figure 4. The micrographs reveal the distribution of lead in the washcoat of a monolithic catalyst after 300 hrs of engine dynamometer testing using ashless oil and a fuel containing 0.06 g Pb/gal and 350 ppm sulfur. This catalyst consisted of a noble metal and a base metal supported on a washcoated cordierite substrate. The washcoat comprised ~15% of the total catalyst weight. The catalyst operating temperature was 649°–704°C. The absorbed electron image depicts the washcoat supported on the cordierite substrate; the 30μ-thick washcoat is the horizontal dark band at the bottom of the micrograph. The large, medium dark areas are cordierite, and the lightest areas are void regions filled with the epoxy used to mount the specimen. The micrograph on the right is the lead x-ray image. The lead appears bright (light); the brighter the area, the greater the lead concentration. It is evident that the lead was distributed throughout the washcoat, and it even penetrated somewhat into the cordierite substrate. The concentration of lead at the interface between the washcoat and the cordierite substrate was especially high. Thus the lead, which was initially deposited at the external surface of the washcoat, easily penetrated the entire depth of the washcoat, and accumulated at the interface between the washcoat and the cordierite substrate. The

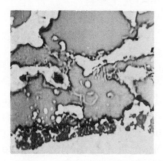 

Figure 4. Distribution of lead in monolithic NO_x catalyst after 300-hr engine test

Left, absorbed electron image; right, x-ray image; fuel, 0.06 g Pb/gal and 350 ppm sulfur; catalyst lead content, 1.1 wt %; and scale, 1 in. = ~68μ

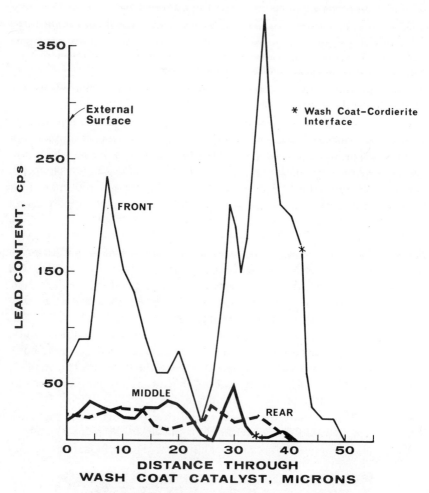

Figure 5. Distribution of lead in monolithic NO_x catalyst after 200-hr engine test

Fuel, 0.06 g Pb/gal and 350 ppm S

cordierite monolithic substrate has substantial macroporosity ($d_p \geq 3\mu$); cordierite itself, however, is a very dense material. Lead can easily move along and into the voids between the washcoated cordierite particles, but it does not penetrate very easily into the cordierite particles.

This same type of lead distribution is depicted more quantitatively in Figure 5 by the EPM line scans for lead in the washcoat of a monolithic catalyst that had been subjected to 200 hrs of engine dynamometer testing using ashless oil and fuel containing 0.06g Pb/gal and 350 ppm sulfur. The scanning was done through the washcoat cross section at three different points along the axis of the monolith, *i.e.* near the front (inlet), middle, and rear (outlet) of the catalyst. This catalyst consisted of a noble metal and a base metal supported on a washcoated cordierite substrate. The washcoat comprised ~15% of the total catalyst weight. Catalyst operating temperature was 649°–704°C. Concentrations of lead at the front of the catalyst were 10–15 times greater than those at the middle and rear. At the front, lead accumulated in two bands. one heavy band near the external surface of the washcoat, and another, even heavier, band at the washcoat cordierite interface.

The apparent accumulation of lead in the voids between the washcoated cordierite particles (Figure 4) is quite interesting. This phenomenon may be caused simply by the agglomeration of metallic lead which is enhanced by the great mobility of this low-melting point metal under conditions of high temperature and a reducing atmosphere. Mooi *et al.* (*1*) demonstrated that lead compounds can react with the cordierite substrate to form lead aluminum silicates by calcining samples of a lead-contaminated HC–CO oxidation catalyst in air at progressively higher temperatures. X-ray analysis of samples calcined at ~1038°C revealed the presence of crystalline $PbAl_2SiO_3$; initially, lead had been present as the sulfate ($PbSO_4$) and oxysulfate ($PbO \cdot PbSO_4$). This finding indicates that chemical interaction between lead compounds and the substrate material can occur during high temperature excursions, *i.e.* periods (normally of short duration) of service during which the catalyst temperature greatly exceeds the normal operating temperature range.

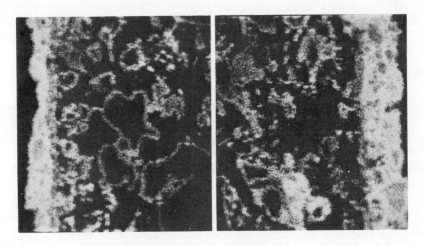

Figure 6. Distribution of washcoat support material on external and internal surfaces of porous ceramic monolithic substrate

X-ray image magnification, ×300; scale, 1 in. = ~40μ

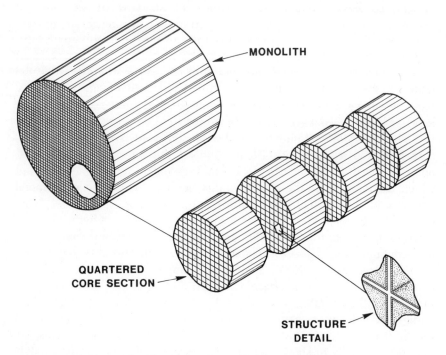

Figure 7. Washcoated monolithic catalyst substrate

The catalyst in Figure 4 was not intentionally exposed to such high temperatures. However, we are also not certain that high temperature excursions did not occur unobserved.

Washcoated Monoliths

The x-ray micrograph of the cross section of a flow channel wall of a washcoated cordierite monolith in Figure 6 reveals how the catalyst support washcoat is distributed on the external surfaces (sides) of the flow channels and throughout the macroporous interior region of the cordierite monolithic substrate. The light areas are the washcoat support material, and the dark areas are cordierite. In the micrograph, the exterior walls (sides) of the channel are vertical. The washcoat on the right wall is noticeably thicker than that on the left wall. It is quite typical that the thickness of the washcoat on the exterior walls varies considerably. The external washcoat is macroporous. The washcoat support material has penetrated into the macroporous structure of the cordierite substrate and has coated its internal surfaces.

The Macro- and Microdistributions of Contaminants on Monolithic NO_x Catalysts

We observed certain trends in the macro- and microdistributions of contaminants on monolithic NO_x catalysts subjected to extended engine dynamometer testing using fully formulated motor oil and certification fuel, i.e. engine oil containing 0.20 wt % calcium, 0.17 wt % zinc, and 0.15 wt % phosphorus and fuel containing 0.03 g lead/gal, 0.005 g phosphorus/gal, and 300 ppm sulfur. Macrodistribution refers to the variation in the gross contaminent concentration in the radial and axial directions. Microdistribution refers to the distribution of contaminants within the

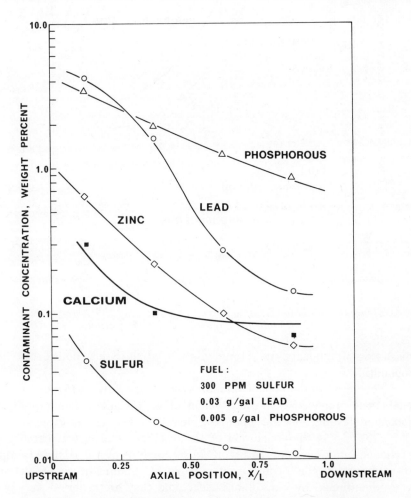

Figure 8. Axial distribution of contaminants on monolithic NO_x catalyst after 40,000-mile dynamometer durability testing

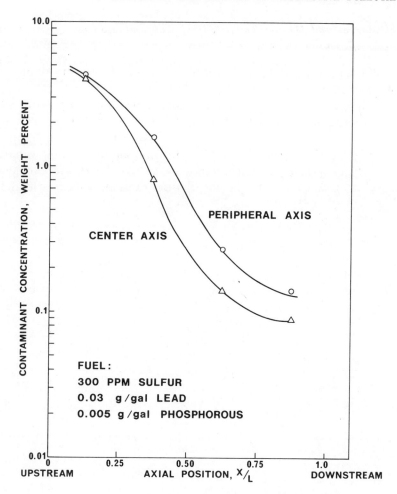

Figure 9. Axial distribution of lead on monolithic NO_x catalyst after 40,000-mile dynamometer durability testing

washcoat of the catalyst at any given axial or radial position in the monolith.

The base metal–noble metal monolithic NO_x catalyst (15 wt % washcoat) was subjected to the equivalent of 40,000 miles of engine dynamometer testing; normal operating temperature of the catalyst was 677°–732°C. At the conclusion of the test, the catalyst was cored and sectioned into quarters (*see* Figure 7) for analysis of the axial and radial distributions of the various contaminants. Contaminant concentrations were determined by x-ray fluorescence, and they are reported as a percentage of the total catalyst weight. Almost all (>90%) of the poisons (Ca, Zn, Pb, P) present on contaminated catalysts were confined, for the most part, to the washcoat. However, contaminant poisons were also present, to some extent, in forms associated with the cordierite substrate. For this reason, contaminant concentrations are reported on the basis of total catalyst weight. The active catalytic metals, on the other hand, were almost completely confined to the washcoat. Since our concern is catalyst degradation (activity loss) caused by contaminant poisoning, we are more interested in the concentrations of contaminants in the washcoat. Since the washcoat of this catalyst comprised ~15% of total catalyst weight, a good approximation of contaminant concentrations in the washcoat can be made by multiplying by six the reported concentrations on the basis of total catalyst weight.

Macrodistribution of Contaminant Poisons. The axial distributions of contaminants are depicted in Figure 8. Contaminant concentrations were greatest at the catalyst inlet, and decreased toward the exit. Note that contaminant concentrations are plotted on a logarithmic scale; thus the variations from inlet to outlet are actually much greater than they appear. The concentrations of lead and phosphorus were especially high (~4 wt %) at the inlet end of the monolith. Large variations in the shape of the sulfur distribution curves were observed for different catalyst samples. The data in Figure 8 are for a peripheral core sample. The axial distributions of calcium, phosphorus, and zinc for peripheral and center axis core samples were very similar, *i.e.* there were no significant radial concentration gradients for these three contaminants. However, the difference in lead concentration in the radial direction was noticeable (Figure 9).

Table III summarizes the data in Figure 8. Lead and zinc (low melting point metals) concentrations were much greater in the front of the catalyst than in the rear (30- and 11-fold respectively). With calcium and phosphorus, the ratio of front to rear concentrations was much less, *i.e.* approximately 4.

Table III. Axial Distribution[a] of Contaminants on NO_x Catalyst after 40,000-mile Dynamometer Testing

Contaminant	Concentration, wt %			Front: Rear Ratio	Contaminant Retention, %
	Front	Rear	Mean		
Phosphorus	3.5	0.85	1.8	4	35
Lead	4.2	0.14	1.3	30	13
Zinc	0.6	0.06	0.24	11	6
Calcium	0.3	0.07	0.13	4	2.6
Carbon	0.06	0.05	0.06	1	—
Sulfur	0.04	0.01	0.02	4	20 ppm

[a] Peripheral core sample.

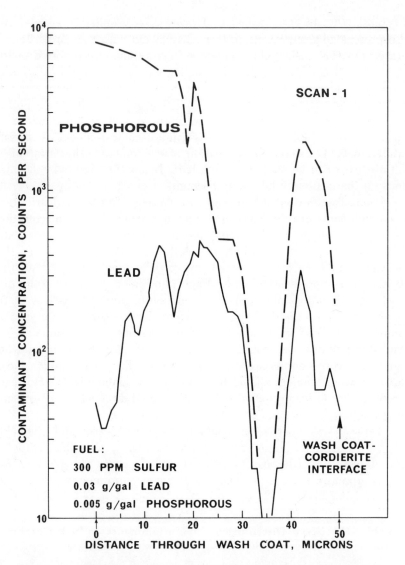

Figure 10. Distribution of contaminants in washcoat of monolithic NO_x catalyst after 40,000-mile dynamometer testing

Retention values for each of the major contaminants are also listed in Table III. Retention is defined as the amount of a particular contaminant present on the catalyst expressed as a fraction of the total amount of that contaminant generated by the consumption of fuel and engine oil. The order of contaminant retention was: phosphorus (35%) > lead (13%) > zinc (6%) > calcium (2.6%) > sulfur (0.002%). The very large retention values for lead and, especially, phosphorus indicate that these two contaminants are highly reactive toward NO_x catalysts.

Shelef et al. (2) reported the following representative contaminant retention values for monolithic noble metal HC–CO oxidation catalysts: lead, 15%; phosphorus, 9%; zinc, 3%; and sulfur, 0.05%. Furthermore, for monolithic noble metal HC–CO catalysts which had been subjected to 30,000 miles of vehicle testing, the ratios of front to rear contaminant poison concentrations were: lead, 7; phosphorus, 16; and zinc, 11. Because NO_x and HC–CO catalysts are normally operated under different ambient conditions, i.e. net reducing vs. net oxidizing atmosphere, it is expected that the nature, distribution, and retention of contaminant poisons will differ for these two types of auto exhaust catalysts.

The external surfaces of the used catalyst, as well as the surfaces of the catalyst container, were usually covered with a grayish-white film that x-ray analysis usually indicated was an amorphous compound consisting of Br, Cl, P, Pb, and S. On occasion, this compound was identified as lead bromide phosphate, $3[Pb_3(PO_4)_2] \cdot PbBr_2$, ~600A crystallite size. Ash and gum deposits were usually found on the upstream (inlet) end of monolithic catalysts; they often caused plugging of a small fraction of the flow channels. Analysis of these deposits revealed that they consisted primarily of calcium, zinc, and phosphorus, i.e. they resulted primarily from the consumption of motor oil.

Microdistribution of Contaminant Poisons. The distributions of lead and phosphorus in the washcoat of a front quarter core section of the catalyst are revealed by the EPM line scans in Figure 10. Throughout the interior portion of the washcoat, the distributions of lead and phosphorus were parallel. This suggests the presence of lead phosphates, e.g.

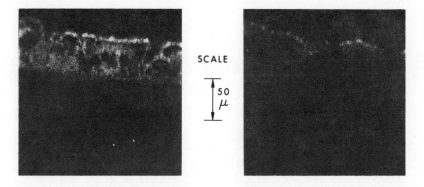

Figure 11. Distribution of contaminant calcium in washcoat of monolithic NO_x catalyst after 40,000-mile dynamometer testing

Left, front core section; right, rear core section; and magnification, ×225

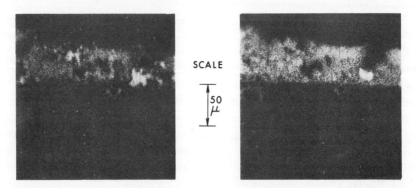

Figure 12. Distribution of contaminant lead and zinc in washcoat of monolithic NO_x catalyst after 40,000-mile dynamometer testing

Front core section; left, *lead;* right, *zinc; and magnification,* ×225

$Pb(PO_3)_2$ and $Pb_3(PO_4)_2$. In the exterior 20% of the washcoat (*i.e.* toward the external washcoat surface), however, the phosphorus concentration was disproportionately greater than the lead concentration. At the external surface of the washcoat, the phosphorus concentration was ~180-fold that of lead. Under the conditions of the exhaust gas atmosphere, phosphorus could be present on the catalyst only in the form of stable phosphate compounds. The high phosphorus concentration in the external surface region of the washcoat indicated that chemical interaction between phosphorus and the washcoat material had occurred to form hydrothermally stable phosphates such as aluminum phosphates, *e.g.* $Al(PO_3)_3$ and $AlPO_4$. X-ray analysis of this catalyst failed to confirm the presence of crystalline aluminum phosphates. It is therefore con-

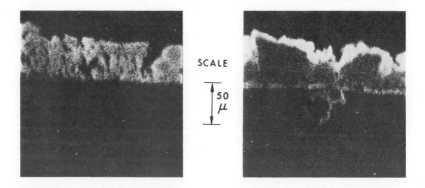

Figure 13. Distribution of contaminant phosphorus in washcoat of monolithic NO_x catalyst after 40,000-mile dynamometer testing

Left, *front core section;* right, *rear core section; and magnification,* ×225

cluded that the aluminum phosphates were present in forms amorphous to x-ray diffraction.

CALCIUM. The x-ray image micrographs of front and rear core sections (Figure 11) reveal the distribution of calcium in the catalyst washcoat. (In Figures 11, 12, and 13, the external surface of the washcoat is at the top of the micrograph.) The presence of a particular contaminant is indicated by the light (bright) areas, and contaminant concentration is directly proportional to image intensity (*i.e.* the brighter the image, the higher the concentration). The x-ray images demonstrate clearly that calcium accumulation in the washcoat was much more extensive in the front core section than in the rear core section. The calcium concentrated heavily in the outer portion of the washcoat (near the external washcoat surface) in both the front and rear core sections.

LEAD AND ZINC. The x-ray images of a front core section (Figure 12) reveal the distributions of lead and zinc in the catalyst washcoat. Lead and zinc appeared to be thoroughly and somewhat randomly distributed in the washcoat; there were also several pockets of very high concentrations of lead and zinc.

PHOSPHORUS. In the x-ray images, there was a very striking difference in the distribution of phosphorus in the catalyst washcoat in the front and rear core sections (Figure 13). In the front (inlet) core section, the phosphorus was distributed uniformly throughout the washcoat, the concentration at the external surface of the washcoat being somewhat higher. In the rear (outlet) core section, the phosphorus was concentrated very heavily at the external surface of the washcoat. There was very little phosphorus in the interior of the washcoat, and the phosphorus concentration at the interface between washcoat and cordierite substrate was greater than that in the washcoat interior. This pattern indicates that phosphorus reactivity toward the catalyst varied considerably between the inlet and outlet ends, perhaps because of changes in exhaust gas composition or differences in the contaminant lead concentrations on the catalyst between these two points. Alternatively, there may be several, gas phase, phosphorus compounds with different reactivities. The most reactive compounds would be removed from the gas phase at the inlet portion of the catalyst.

One mechanism for the interaction between phosphorus and catalyst would be: adsorption of P_4O_{10} and/or P_4O_6, followed by their reaction on the surface with adsorbed H_2O to form H_3PO_4, then reaction between H_3PO_4 and the catalyst to form hydrothermally stable phosphate compounds. The x-ray images revealed high concentrations of both calcium and phosphorus at the external surface of the washcoat. This suggests the presence of calcium phosphates, *e.g.* $Ca(PO_3)_2$, $Ca_3(PO_4)_2$, and $Ca_2P_2O_7$.

The x-ray images presented in Figures 11, 12, and 13 indicate that the various contaminants were confined largely to the catalyst washcoat. That is, the contaminants did not penetrate to any significant extent into the cordierite substrate. The washcoat of the catalyst pictured in these micrographs was 30–40μ thick.

Electron Spectroscopy Chemical Analysis (ESCA)

In a preliminary ESCA analysis of the catalyst surface, sulfur was not detected and phosphorus was present in its highest valence state, *i.e.* P^{+5}. The front-to-rear ratios of the surface concentrations of lead and phosphorus were 6 and 1.5 respectively, whereas the front-to-rear ratios of their bulk concentrations were 30 and 4 respectively.

Acknowledgment

The author expresses his gratitude to his colleague Howard D. Simpson for his contributions to the thermodynamic model and calculations which were used to predict catalyst metal sulfiding in a synthetic exhaust gas atmosphere.

Literature Cited

1. Mooi, J., Kuebrich, J. P., Johnson, M. F. L., Chloupek, F. J., "Modes of Deactivation of Exhaust Purification Catalysts," Amer. Petrol. Inst., Mid-year Mtg., 38th, Div. Refining, Philadelphia, May, 1973.
2. Shelef, M., Dalla Betta, R. A., Larson, J. A., Otto, K., Yao, H. C., "Poisoning of Monolithic Noble Metal Oxidation Catalysts in Automobile Exhaust Environment," Symp. Catal. Poisoning, Amer. Inst. Chem. Eng., Nat. Mtg., 74th, New Orleans, March, 1973.

RECEIVED June 7, 1974.

9

The Chemistry of Degradation in Automotive Emission Control Catalysts

RICHARD L. KLIMISCH, JERRY C. SUMMERS, and
JAMES C. SCHLATTER

General Motors Research Laboratories, Warren, Mich.

The parameters that affect the degradation of supported platinum and palladium automotive exhaust catalysts are investigated. The study includes the effects of temperature, poison concentration, and bed volume on the lifetime of the catalyst. Thermal damage primarily affects noble metal surface area. Measurements of specific metal area and catalytic activity reveal that supported palladium is more thermally stable than platinum. On the other hand, platinum is more resistant to poisoning than palladium. Electron microprobe examinations of poisoned catalyst pellets reveal that the contaminants accumulate almost exclusively near the skin of the pellet as lead sulfate and lead phosphate. It is possible to regenerate these poisoned catalysts by redistributing the contaminants throughout the pellet.

Catalyst durability is of primary concern in the use of catalysts for control of automotive emissions. There are many reasons for this intense interest. An obvious one is that the federal emissions standards specify 50,000 miles or five years, whichever occurs first. Economic considerations such as material supplies, balance of payments, and costs of catalyst replacement put a premium on developing catalysts that are resistant to degradation. There are two primary causes for loss of catalytic activity— high temperature and contamination. Exposure to elevated temperatures causes physical deterioration of supported metal catalysts; the effects include loss of active surface area and, in extreme cases, destruction of the support itself. Certain trace components in the fuel can also decrease catalyst efficiency; this chemical deterioration is not necessarily associated with high temperatures.

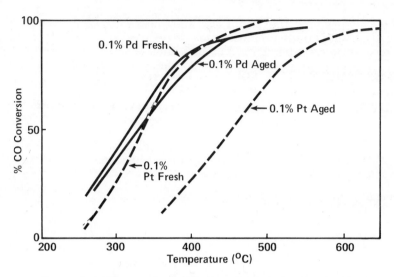

Figure 1. Effect of thermal aging on CO oxidation activity
Thermal aging: 70 hrs at 900°C

Catalysts in automotive exhaust systems are subjected to a wide range of environments, and both chemical and physical deterioration are likely to occur. In many cases, the two effects are inseparable. Nevertheless, we must understand both in order to minimize their consequences. As a first step toward such an understanding, we observed catalyst characteristics following exposure to damaging conditions. As happens in most experimental studies, the initial work gave rise to many questions, some of which are currently under active investigation in our laboratories. In spite of these questions, we feel it is appropriate to present our findings at this time in the hope that they will provoke further research into this challenging and important area of catalysis.

Experimental

Supported platinum and palladium catalysts were prepared by soaking a preformed alumina support (Kaiser KC/SAF gel-derived alumina, 250 m²/g, ⅛-inch spheres) in concentrated aqueous solutions of the appropriate metal chloride. The catalysts were calcined at 600°C in air before use. Specific metal surface areas were measured by titration of chemisorbed oxygen with hydrogen (1, 2). Prior to the adsorption measurement, each sample was reduced 1 hr in H_2 at 500°C, evacuated 1 hr at 500°C, and exposed overnight to O_2 at room temperature. Elemental distributions in the catalyst pellets were determined by electron microprobe.

The effects of contaminants on catalytic activity were determined by subjecting samples to converter usage on an engine operated with fuels containing varying amounts of lead and phosphorus. Catalyst efficiencies were measured during the poisoning procedure (at 15-min intervals) and afterward in the laboratory. Thermal aging experiments were conducted in a laboratory atmosphere. In the poisoning tests, a rather large amount (980–4250 cm^3) of catalyst was used to treat exhaust from a full-size (350 cid) engine on a test stand. For laboratory activity measurements, 10–20 cm^3 of catalyst were used to oxidize carbon monoxide and hydrocarbons in a stream blended from cylinder gases to resemble exhaust gas. In the engine tests, conversion was measured continuously at the catalyst bed temperature reached during steady operation of the engine, typically 550°C. In the laboratory tests, conversion was recorded as a function of temperature. The feedstream for the laboratory tests consisted of 2% CO, 0.5% propylene, 2.5% O_2, 10% H_2O, and 10% CO_2 in a nitrogen atmosphere at a gas hourly space velocity (GHSV) of 85,000/hr (3).

Results and Discussion

Thermal Stability. Our initial work on catalyst degradation processes was designed to answer qualitatively several simple yet vital questions. This report is organized similarly. The question naturally arises whether platinum or palladium is best for automotive emission control. One aspect

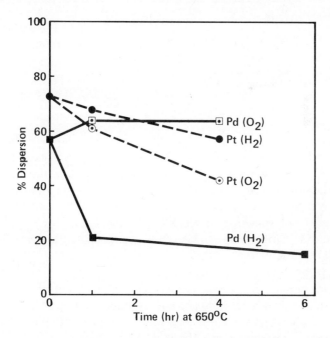

Figure 2. Dispersion stability of alumina-supported (0.4 wt %) platinum and palladium in H_2 and in O_2 at 650°C

Stability: $Pd(O_2) > Pt(H_2) > Pt(O_2) >> Pd(H_2)$

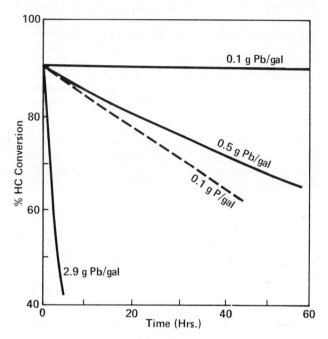

Figure 3. Effect of lead level in fuel on exhaust catalyst deterioration

Catalyst: 1000 cm^3 0.1% Pt on Al_2O_3 at 550°C; engine: 350 cid at 1800 rpm, 14-in. Hg manifold vacuum ($\cong$ 45 mph), A/F = 16; and inlet composition: 0.06% HC, 0.3% CO, 2.0% O_2, and 0.1% NO

of this question involves thermal stability and, in particular, the relative thermal stabilities as a function of environment. Our data have consistently demonstrated that palladium catalysts maintain their activity at high temperature in oxidizing atmospheres much better than their platinum counterparts. For example, a 70-hr thermal treatment in static air at 900°C did not appear to affect the oxidation activity of a 0.1% Pd on Al_2O_3 catalyst (Figure 1). The corresponding platinum catalyst was substantially deactivated by the same thermal treatment.

It was obviously of interest to determine if these activity changes correspond to changes in the active surface areas of these catalysts. Because contaminants accumulate on catalysts exposed to auto exhaust, adsorption values may not be valid indicators of metal particle size for such samples. Therefore, our thermal studies were conducted in pure gases in order to avoid complications from exhaust contaminants. The data from sintering experiments at 650°C in hydrogen and in oxygen atmospheres (with flowing gases) are plotted in Figure 2. The findings for the two metals were quite different. The increased sintering of alumina-supported platinum in an oxidizing atmosphere, compared with that in reducing atmosphere, has been reported (4), and it is consistent with the activity loss for platinum depicted in Figure 1. Interestingly, the findings for palladium in these experiments were just the opposite and were also considerably more dramatic. Thus, whereas an oxidizing atmosphere at 650°C apparently did not sinter the palladium at all, just 1 hr in hydrogen at the same temperature decreased the palladium titration uptake by 60%.

It is beyond the scope of this report to attempt to explain the data in Figure 2. A recent theoretical analysis by Wynblatt and Gjostein (5) predicts the sequence observed in these experiments. We have x-ray evidence that a substantial portion of the palladium is present as the oxide after exposure to an oxidizing environment. Presumably, oxide formation is expected to stabilize high dispersions (5). The apparent rapid sintering of palladium in hydrogen is certainly intriguing and is under investigation. However, for converter application, the superior thermal stability of palladium in oxidizing environments is unmistakable.

Catalyst Poisoning. Our investigations of catalyst poisoning revealed that the major factors determining catalyst lifetime in a poisoning environment are fuel contaminant level and catalyst volume. The effect of lead level in the fuel is depicted in Figure 3. These data were obtained in accelerated poisoning tests using a relatively small catalyst volume

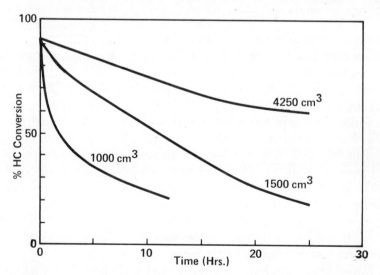

Figure 4. Effect of catalyst volume on resistance to poisoning

Fuel: 2.9 g Pb/gal; catalyst, engine, and inlet composition: same as in
Figure 3

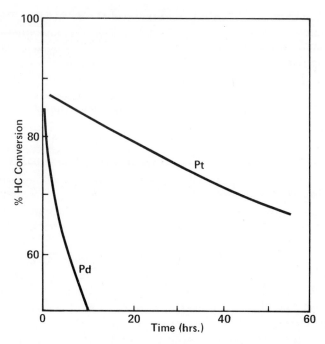

Figure 5. Resistance of platinum and palladium to poisoning by 0.4 g Pb/gal fuel

Catalyst: 1000 cm^3 0.1% Pt or 0.1% Pd on Al_2O_3 at 550°C; engine and inlet composition: same as in Figure 3

(1000 cm^3) to treat the exhaust from a 350 cid engine. In the time scale of these experiments (ca. 60 hr), 0.1 g Pb/gal appeared to have little effect on the catalyst. On the other hand, activity loss was severe in 60 hrs with 0.5 g Pb/gal, and similar activity loss occurred in only 1 hr with fully leaded gasoline, i.e. 2.9 g Pb/gal. These findings are consistent with earlier studies of the effect of fuel lead content on catalyst lifetime (6, 7). For comparison, the data for phosphorus poisoning are also presented in Figure 3; they suggest that phosphorus and lead have about equal potency on an atom-for-atom basis.

The effect of a fourfold change in catalyst volume on catalyst deterioration is depicted in Figure 4. The smallest converter (1000 cm^3) had a rapid initial loss in activity followed by a milder loss rate. The milder loss rate paralleled the rate of activity loss of the larger converters. The larger converter was clearly better for poison tolerance.

The effect of catalyst type is depicted in Figure 5. Platinum was much more resistant to lead poisoning than palladium. Giacomazzi and Holmfeld (8) reached the same conclusion with vehicle tests using lower levels of lead in the fuel. It is probably not surprising that platinum was more resistant to poisoning than palladium, and the dramatic difference at the high poison level of these experiments (0.4 g Pb/gal) suggests that poisoning is not a random process since it reflects the inherent reactivity differences between the metals.

Having made these qualitative observations on the parameters that control poisoning, we subjected the catalysts to detailed analysis. As in our earlier poisoning study (3), electron microprobe analysis revealed that the poison accumulated almost exclusively at the outer edge of the catalyst pellet. The relative concentrations of lead and sulfur, as well as x-ray diffraction, suggest that the lead was primarily in the form of lead sulfate (3, 6). The findings from a typical microprobe analysis of a poisoned catalyst are presented in Figure 6.

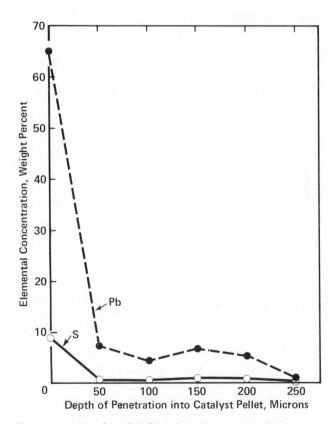

Figure 6. Lead and sulfur distribution in a lead-poisoned catalyst pellet

Poisoning: 100-hr exposure to exhaust from engine using fuel with 0.5 g Pb/gal

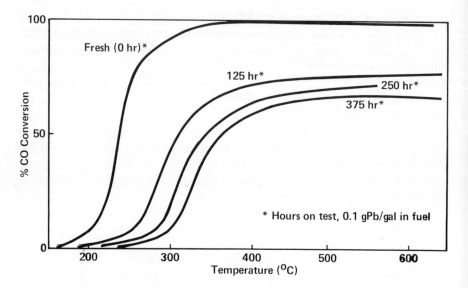

Figure 7. Laboratory evaluation of a lead-poisoned Pt–Al_2O_3 catalyst
GHSV: 90,000/hr; lead accumulation: 0.1 g Pb/gal; converter: 2300 cm^3 radial flow at 590°C; and engine: same as in Figure 3

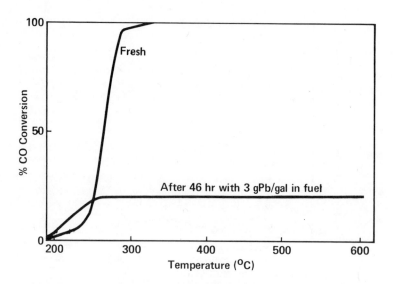

Figure 8. Laboratory evaluation of a lead-poisoned 0.1% Pt–Al_2O_3 catalyst
GHSV: 85,000/hr

The very high levels of lead at the outer region of the pellets had a striking effect on the laboratory activity profiles obtained for the poisoned catalysts. This was most clear for catalysts that were poisoned at the lowest lead level, 0.1 g Pb/gal. The activity profiles for samples run for various times are presented in Figure 7. It is not surprising that a higher temperature was required to attain a given conversion as the catalyst became progressively more contaminated. It is especially interesting that the conversion did not reach completion. The poisoned catalysts had fairly normal light-off characteristics followed by severely limited maximum conversion. The conversion was limited to an intermediate value which apparently depended on the degree of catalyst exposure to lead. An activity profile for a more drastically poisoned catalyst is given in Figure 8.

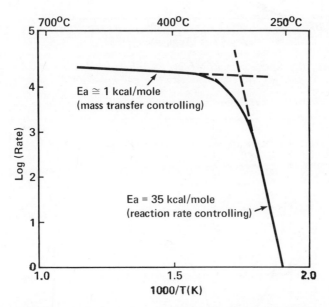

Figure 9. Arrhenius plot for a lead-poisoned catalyst after 250-hr accumulation using fuel with 0.1 g Pb/gal

The Arrhenius plot for a poisoned catalyst (Figure 9) reveals two distinct regions: a low temperature region characterized by normal activation energy and a high temperature region characterized by a very low activation energy. Since low activation energies are characteristic of diffusion processes, and since the microprobe clearly demonstrated that the contaminants were concentrated near the pellet surface, it is logical to propose that this inactive area limits the conversion at high temperatures by forcing the reactants to penetrate the catalyst pellet in order to react.

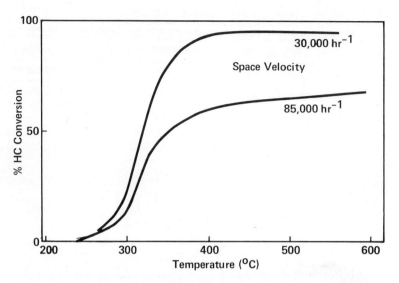

Figure 10. Space velocity effect: laboratory evaluation of a Pt–Al_2O_3 catalyst after exposure to 0.1 g Pb/gal for 375 hrs

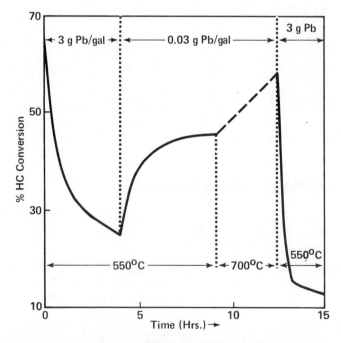

Figure 11. Cumulative effect of successive lead poisoning on a Pt–Al_2O_3 exhaust catalyst

Catalyst, engine, and inlet composition: same as in Figure 3

One practical aspect of this effect is illustrated by the laboratory reactivity profiles (Figure 10) for a poisoned catalyst at two space velocities. At GHSV = 30,000/hr, no limit on conversion was apparent; however, at GHSV = 85,000/hr, the catalyst attained a maximum conversion of only 60%. Thus, in contrast to earlier reports (9, 10), lead levels as low as 0.1 g Pb/gal did damage automotive catalysts, and, in particular, they would be expected to limit seriously conversion at high-speed operation. Since the FTP vehicle test emphasizes low-speed operation, and since vehicle tests have considerable variability, the effect will probably not be apparent until more data are analyzed. These results are consistent with the earlier studies of converter size (Figure 4), and they provide considerable insight into the poisoning process.

The ability of catalysts to recover after poisoning by lead has been noted (9, 11). The platinum catalyst recovered part of its activity when lead was removed from the fuel (*see* Figure 11). Furthermore, the regen-

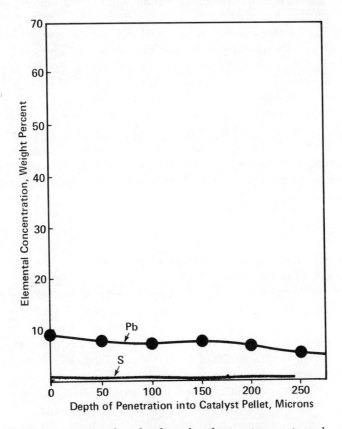

Figure 12. Lead and sulfur distribution in a poisoned pellet regenerated in air 90 hrs at 900°C

eration process continued and presumably was accelerated at higher temperature. Microprobe analysis of a regenerated sample (Figure 12) revealed that the sharp demarcation of contaminants at the edge of the pellet was transformed into a distribution extending well into the pellet interior. In addition, the lead no longer was present as lead sulfate since the sulfur had disappeared. We conclude that the primary process in activity regeneration involves destruction of the lead sulfate-poisoned layer and redistribution of the lead throughout the interior of the pellet. Since the lead was not actually removed during regeneration (as determined by chemical analysis), it is reasonable to expect that a second exposure will be more detrimental than the initial exposure (Figure 11). Volatilization of the lead did not occur until higher temperatures ($> 900°C$). Some of the initial activity recovery depicted in Figure 11 may have resulted from the desorption of halogens (*12*).

Although thermally aged catalysts tended to lose low temperature activity while still maintaining conversion capability at high temperature (Figure 1), the characteristic changes of poisoned catalysts were just the opposite. The conversion lid on poisoned catalysts, discussed above, was attributed to a diffusion effect caused by poison buildup near the pellet skin. The shift in activity characteristic of thermally aged catalysts can be attributed to loss of platinum sites. The same kind of change occurred in conversion–temperature plots when CO concentration was increased. (Carbon monoxide essentially acts to eliminate sites for its own oxidation.) Thus, the conversion efficiency of an unpoisoned platinum or palladium catalyst is primarily a function of the ratio of CO concentration to the metal surface area. The change in shape of laboratory activity profiles can perhaps be used as a diagnostic technique for catalyst deterioration problems.

Summary

Several qualitative conclusions can be drawn from the preliminary work described in this paper.

(a) High-temperature oxidizing atmospheres decreased the activity and surface area of platinum catalysts more than those of palladium catalysts.

(b) Platinum was more resistant than palladium to poisoning.

(c) Lowering the fuel lead levels (0.1 *vs.* 2.9 g Pb/gal) and increasing the converter volume increased durability.

(d) Poisons accumulated near the outer surface of the catalyst pellets and limited conversion through a diffusion resistance.

(e) Thermal treatments could partially regenerate exhaust catalysts through a redistribution of contaminants.

(f) The activity profile of a used catalyst depended on whether the sample was deactivated by overheating or by contamination. Most of these are the subject of continuing research in our laboratories.

Literature Cited

1. Boudart, M., Benson, J. E., J. Catal. (1965) **4,** 704.
2. Benson, J. E., Hwang, H. S., Boudart, M., J. Catal. (1973) **30,** 146.
3. Gallopoulos, N. E., Summers, J. C., Klimisch, R. L., "Effects of Engine Oil Composition on the Activity of Exhaust Emissions Oxidation Catalysts," Soc. Automot. Eng., Automot. Eng. Mtg., Detroit, May, 1973, paper **730598**.
4. Somorjai, G. A., Powell, R. E., Montgomery, P. W., Jura, G., in "Small Angle X-Ray Scattering," H. Brumberger, Ed., pp. 449–466, Gordon & Breach, New York, 1967.
5. Wynblatt, P., Gjostein, N. A., Progr. Solid State Chem. (1974) **9,** in press.
6. Yarrington, R. M., Bambrick, W. E., J. Air Pollut. Control Ass. (1970) **20,** 398.
7. Weaver, E. E., "Effects of Tetraethyl Lead on Catalyst Life and Efficiency in Customer Type Vehicle Operation," Soc. Automot. Eng., Automot. Eng. Congr., Detroit, January, 1969, paper **690016**.
8. Giacomazzi, R. A., Holmfeld, M. F., "The Effect of Lead, Sulfur, and Phosphorus on the Deterioration of Two Bead-Type Oxidizing Catalysts," Soc. Automot. Eng., Automot. Eng. Mtg., Detroit, May, 1973, paper **730595**.
9. Neal, A. H., Wigg, E. E., Holt, E. L., "Fuel Effects on Oxidation Catalysts and Catalyst-Equipped Vehicles," Soc. Automot. Eng., Automot. Eng. Mtg., Detroit, May, 1973, paper **730593**.
10. Hetrick, S. S., Hills, F. J., "Fuel Lead and Sulfur Effects on Aging of Exhaust Emission Control Catalysts," Soc. Automot. Eng., Automot. Eng. Mtg., Detroit, May, 1973, paper **730596**.
11. Gagliardi, J. C., Smith, C. S., Weaver, E. E., "Effect of Fuel and Oil Additives on Catalytic Converters," Amer. Petrol. Inst. Prepr. **63–72,** 1972.
12. Holt, E. L., Wigg, E. E., Neal, A. H., "Fuel Effects on Oxidation Catalysts and Oxidation Catalyst Systems—II," Soc. Automot. Eng., Automot. Eng. Congr., Detroit, January, 1974, paper **740248**.

RECEIVED May 28, 1974.

10

The Optimum Distribution of Catalytic Material on Support Layers in Automotive Catalysis

JAMES WEI and E. ROBERT BECKER

Department of Chemical Engineering, University of Delaware, Newark, Del. 19711

> *Classical analysis has demonstrated that a given quantity of active material should be deposited over the thinnest layer possible in order to minimize diffusion limitations in the porous support. This conclusion may be invalid for automotive catalysis. Carbon monoxide oxidation over platinum exhibits negative order kinetics so that a drop in CO concentration toward the interior of a porous layer can increase the reaction rate and increase the effectiveness factor to above one. The relative advantage of a thin catalytic layer is further reduced when one considers its greater vulnerability to attrition and to the deposition of poisons. This analysis is demonstrated, and it is concluded that a thick layer may have advantages over a thin layer.*

The most effective way to utilize a given quantity of catalytic material is to deposit it on a layer of porous support. The classical Thiele analysis demonstrated that, for a first order reaction, it is preferable to concentrate the active ingredients in a thin layer to minimize diffusion effects; this conclusion remains valid for any positive order kinetics. This diffusion effect causes a decline in reactant concentration toward the interior of a porous catalytic layer, leading to a decline in reaction rate in the interior. When the Thiele modulus is sufficiently large, such as when the reaction rate is fast and when diffusion through a porous layer is slow, only a very thin layer on the exterior is contributing to the reaction rate.

Thus, recent automotive catalyst development tends toward depositing platinum in a thin washcoat of alumina on low-surface-area, ceramic monoliths or in egg shell layers on the exterior of pellets.

However, the kinetics of CO and hydrocarbon oxidation over platinum has a negative first order dependence on CO concentration for most of the ranges of temperature and concentration of interest to automotive catalysis. Voltz et al. (1) demonstrated that, under a total pressure of 1 atm. at 400°–800°F, the kinetics depend inversely on CO concentration from 0.2% to 4%. Therefore, when there are diffusion effects, a decline in CO concentration toward the interior of a porous catalytic layer would lead to an increase in reaction rates in the interior. Diffusion then has a beneficial rather than a detrimental effect on overall kinetic rates, and a thick catalytic layer may effect a higher conversion than a thin layer under identical ambient conditions.

The advantages of a thick catalytic layer may be even greater when poisoning effects are considered. Even when the concentration of lead in gasoline is reduced to 0.05 g/gal, more than 0.5 lb of lead passes over the catalysts during 50,000 miles of driving. Recent data indicate that lead deposition is highly concentrated on a thin exterior layer of the catalyst. Under these conditions, platinum near the surface is easily covered by lead deposition and rendered ineffective while platinum in the interior is better protected from lead deposition.

The geometric configurations of the catalysts may be ranked by their effectiveness factors, but a better measure is their CO conversion efficiency under identical inlet concentrations, space velocities, and temperatures. Catalyst deterioration caused by poison deposition and thermal damage should also be considered. These factors should be quantified in order to ascertain the optimum thickness of catalytic layers.

The Effectiveness Factor

In automotive catalysis, as in any heterogeneous reaction, both internal and external mass transport are significant. The effect of mass transport on reactions of positive order and Langmuir–Hinshelwood kinetics have been discussed by Satterfield (2), Roberts and Satterfield (3), and Petersen (4). With CO oxidation over platinum catalyst, the reaction rate exhibits a negative first order dependence on the reactant concentration of the form:

$$\text{rate} = \frac{K \cdot C_{O_2} \cdot C}{(1 + K_a C)^2} = \frac{K_r C}{(1 + K_a C)^2} \quad (1)$$

where K_r is the reaction rate parameter (1/sec) $= KC_{O_2}$, K_a is the absorption rate constant (1/mole % CO), C is the CO concentration (mole %), and C_{O_2} is the O_2 concentration (mole %).

This kinetic form of the rate was recently reported by Voltz et al. (1). When oxygen is present in excess, its concentration can be absorbed

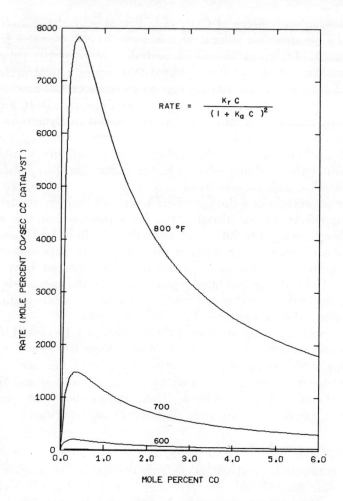

Figure 1. Reaction rate vs. concentration for CO oxidation on platinum

in a parameter K_r. The shape of the curves (Figure 1) is similar to that for exothermic first order reactions in porous catalysts where the internal temperature gradients lead to unusual behavior (5).

Flat plate geometry, corresponding to layer deposition of porous medium on monoliths, was selected as the basis for the analysis of diffusion effects. Other catalyst geometries would yield similar results and conclusions. A component mass balance for CO in the catalyst pore under isothermal conditions yields:

$$D_e \frac{d^2C}{dx^2} = \frac{K_r C}{(1 + K_a C)^2} \qquad (2)$$

with the boundary conditions for pore mouth: $C = C_s$ at $x = l$, and for pore end: $dC/dx = 0$ at $x = 0$.

When oxygen is not in excess, it is necessary to include the parameter E introduced by Roberts and Satterfield (3)

$$E = \frac{2 D_{O_2} C_{O_2,s}}{D_{CO} C_s} - 1$$

The stoichiometry of the reaction requires that:

$$\frac{C_{O_2}}{C_{O_2,s}} = \frac{E + C/C_s}{E + 1}$$

Substitution into Equation 2 gives:

$$D_e \frac{d^2C}{dx^2} = \frac{K \cdot C_{O_2,s} C \left[\frac{E + C/C_s}{E + 1} \right]}{(1 + K_a C)^2} = \frac{K_r \cdot C}{(1 + K_a C)^2} \cdot \frac{(E + C/C_s)}{(E + 1)} \quad (2a)$$

For the rest of this paper, only the case $E \to \infty$ will be considered.

By introducing $\bar{x} = x/l$ dimensionless pore length, $y = C/C_s$ dimensionless concentration, $\phi = l\sqrt{K_r/D_e}$ Thiele modulus, and $\gamma = 1/K_a C_s$ adsorption parameter, Equation 2 reduces to:

$$\frac{d^2y}{d\bar{x}^2} = \phi^2 \gamma^2 \frac{y}{(\gamma + y)^2} \quad (3)$$

where $y = 1$ at $\bar{x} = 1$ and $dy/d\bar{x} = 0$ at $\bar{x} = 0$.

By defining the effectiveness factor as the measure of diffusional effect in the usual manner:

$$\eta = \frac{\text{observed reaction rate}}{\text{reaction rate without diffusion}}$$

$$\eta = \frac{D_e \, dC/dx \, |_{x = l}}{l \, K_r \, C_s/(1 + K_a C_s)^2} \quad (4)$$

In terms of the reduced variables, Equation 4 reduces to:

$$\eta = \frac{(\gamma + 1)^2}{\gamma^2 \phi^2} \frac{dy}{d\bar{x}} \bigg|_{\bar{x} = 1} \quad (5)$$

Equation 3 may be solved by letting $p = dy/d\bar{x}$ to give

$$p \frac{dp}{dy} = \gamma^2 \phi^2 \frac{y}{(\gamma + y)^2} \quad (6)$$

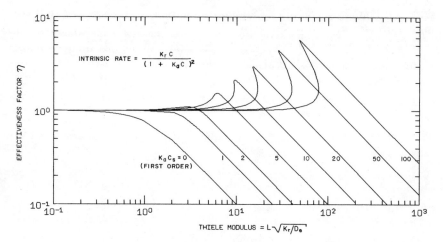

Figure 2. Effectiveness factor vs. Thiele modulus for CO oxidation on platinum

and integrating from $\bar{x} = 0$ where $y = y_0$ and $p = 0$

$$p^2 = \left(\frac{dy}{d\bar{x}}\right)^2 = 2\gamma^2\phi^2\left[\frac{(y_0 - y)\gamma}{(y + \gamma)(y_0 + \gamma)} + \ln\frac{(y + \gamma)}{(y_0 + \gamma)}\right]$$

$$\frac{dy}{d\bar{x}} = \gamma\phi\sqrt{2\left[\frac{(y_0 - y)\gamma}{(y + \gamma)(y_0 + \gamma)} + \ln\frac{(y + \gamma)}{(y_0 + \gamma)}\right]} \quad (7)$$

We note here that

$$\left.\frac{dy}{d\bar{x}}\right|_{\bar{x}=1} = \gamma\phi\sqrt{2\left[\frac{(y_0 - 1)\gamma}{(1 + \gamma)(y_0 + \gamma)} + \ln\frac{(1 + \gamma)}{(y_0 + \gamma)}\right]} \quad (8)$$

and, for $y_0 \ll 1$:

$$\left.\frac{dy}{d\bar{x}}\right|_{\bar{x}=1} = \gamma\phi\sqrt{2\left[\ln\frac{(1 + \gamma)}{\gamma} - \frac{1}{(\gamma + 1)}\right]} \quad (9)$$

Equation 9 combined with Equation 5 gives the asymptotic behavior of the effectiveness factor for large values of the Thiele modulus. Other values for the effectiveness factor are obtained by numerical integration of Equation 7 to yield the desired concentration profile:

$$\int_0^{\bar{x}} d\rho = \frac{1}{\gamma\phi}\int_{y_0}^y \frac{d\sigma}{\sqrt{2\left[\ln\frac{\sigma + \gamma}{y_0 + \gamma} - \frac{(\sigma - y_0)\gamma}{(\sigma + \gamma)(y_0 + \gamma)}\right]}} \quad (10)$$

In this problem with split boundary conditions, the values of y_o and (dy/dx) at $\bar{x} = 1$ are unknown. One must make an initial guess for y_o to start the integration, but, upon arriving at $y = 1$, the value of x will be at some value ξ other than 1. A trial and error approach can be avoided by assuming an interim value of ϕ. The integration can be started by choosing values for γ and y_o, as well as a provisional value, $\phi = \phi_p$. Integration of Equation 10 gives:

$$\xi = \frac{1}{\gamma \phi_p} \int_{y_o}^{1} \frac{d\sigma}{\sqrt{2\left[\ln\left(\frac{\sigma+\gamma}{y_o+\gamma}\right) - \frac{(\sigma - y_o)\gamma}{(\sigma+\gamma)(y_o+\gamma)}\right]}}$$

which is the solution where $\phi = \xi \phi_p$. The value of η can be computed by combining Equations 5 and 8. When $y_o \neq 0$, the existence of the

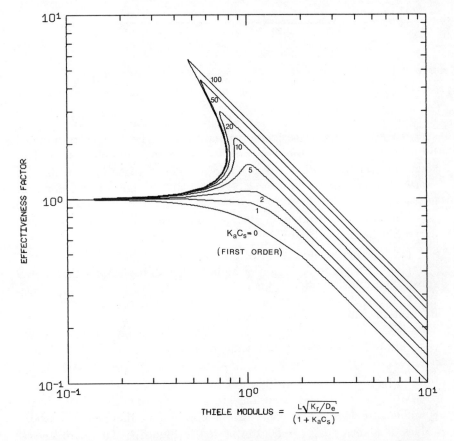

Figure 3. *Effectiveness factor vs. Thiele modulus for CO oxidation on platinum*

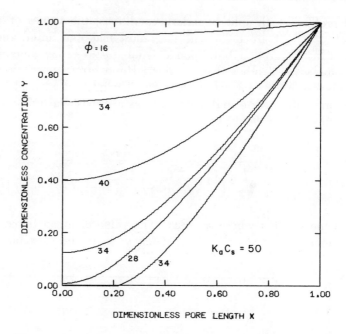

Figure 4. Concentration profiles for changing Thiele modulus

integral is guaranteed, and the numerical procedure converges when successively smaller space increments are used. The singularity induced by $y_o = 0$ is given in the Appendix. A Simpsons rule integration gives sufficient accuracy within a reasonable number of space increments, and more accurate predictor–corrector formulae do not offer significant advantage.

The curves generated in this way are plotted in Figure 2. Each curve can be divided into:

(a) the horizontal region of low ϕ where $\eta = 1$,

(b) the asymptotic region of high ϕ where η is given by Equation 9 substituted into Equation 5, and

(c) the curved region of intermediate ϕ where η is computed by integrating Equation 10.

The 15–20 points computed for each curved region were especially concentrated near the maximum value of η. Defining the Thiele modulus: $\phi = \{L\sqrt{K_r/D_e}\}/\{1 + K_aC_s\}$ and plotting it vs. η produces the compressed family of curves in Figure 3.

Multiple steady-state solutions are observed when the value of K_aC_s is sufficiently high. The concentration profiles in the pore at various values of ϕ and at $K_aC_s = 50$ are plotted in Figure 4. There are several interesting features. For substantial inhibition effects, *i.e.* at large values of K_a or reactant concentration, the unity effectiveness factor is extended to larger values for the Thiele modulus than for the corresponding first order reaction ($K_a C_s = 0$). This is entirely expected since the intrinsic kinetics is proportional to $(1 + K_a C)^{-2}$. Then a region of improved efficiency appears which corresponds to those situations where the concentration drop in the pore results in a reaction rate greater than the rate without diffusion. The effectiveness factor drops below unity only when a very substantial portion of the active surface is not utilized. A relatively small hysteresis can be expected from an increase in temperature from the horizontal to the asymptotic region, followed by an equivalent decrease.

The curves in Figures 2 and 3 should be compared with those of Roberts and Satterfield (3). We calculated the case where $E = \infty$, and

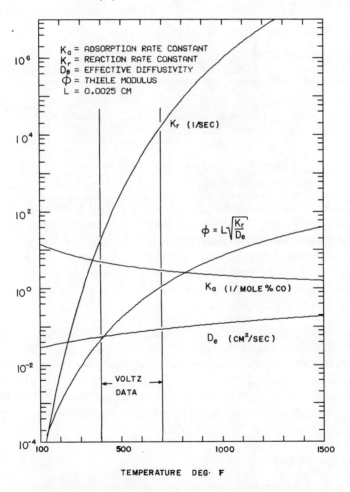

Figure 5. Temperature dependence of parameters

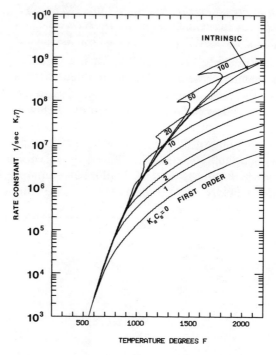

Figure 6. *Observed rate constant vs. temperature*

our Thiele modulus does not include the absorption constant K_a. They calculated for $E = 0$, 1, and 10, and their Thiele modulus did include K_a. The curves are similar in appearance.

A summary of the Voltz data and the Thiele modulus is presented in Figure 5. The observed rate parameters are plotted in Figure 6.

The non-isothermal version of analysis produces the parameter β (2)

$$\beta = \frac{C_s(-\Delta H)D_e}{\lambda T_s}$$

Even under the most extreme conditions likely to be encountered in automobile exhaust gas, $\beta \leq 0.02$. Thus, there are no significant intraparticle heat effects in this system, and isothermal treatment of diffusion is adequate. However, the temperature difference between the gas and the solid is expected to be much higher.

Conversion of CO

The real concern in automotive catalysis is the removal of the pollutants such as CO. For a given kinetic expression and quantity of active catalytic ingredient at a given space velocity through a given reactor, the efficiency of CO removal depends on the inlet temperature and CO concentration as well as on the previously discussed diffusion effects. The rate of external mass transport, the heat generated in the CO conversion, the decline in selfpoisoning of CO oxidation as conversion proceeds, and the deposition of lead on exterior layers all play a role. To simplify these complexities, we present here an initial analysis which ignores these four factors.

Consider an isothermal monolithic reactor containing a fixed quantity of noble metal or base metal dispersed over a support layer of thickness l. The space inside the reactor is divided into the porous catalytic layer, the low porosity wall, and the void space. The three volume fractions sum to unity.

$$\varepsilon_c + \varepsilon_w + \varepsilon_g = 1$$

If the reactor behaves as a piston flow reactor, then a mass balance for CO yields

$$v \frac{dC}{dz} = -\frac{\varepsilon_c}{\varepsilon_g}(r) \qquad (11)$$

where v is the linear velocity of the gas, r is the reaction rate, and z is the distance along the reactor length. Equation 11 should be integrated from $z = 0$ to L and $C = C_o$ to C_L.

$$\int_{C_0}^{C_L} \frac{dC}{r} = -\frac{L}{v}\frac{\varepsilon_c}{\varepsilon_g} \qquad (12)$$

For a first order reaction, in the presence of diffusion effects,

$$r = K_1 \, \eta \, C$$

where $\eta = \tanh \phi_1/\phi_1$, the effectiveness factor and $\phi_1 = l\sqrt{K_1/D_e}$. Substitution into Equation 12 yields the familiar

$$y_L = \exp\left(-\frac{\varepsilon_c}{\varepsilon_g} K_1 \eta \tau\right) \qquad (13)$$

where τ is L/v, the residence time, and y_L is the fraction of CO remaining or $(1 - \text{conversion})$.

For the CO oxidation kinetics over platinum,

$$r = \frac{K_r \eta C}{(1 + K_a C)^2}$$

Integration yields

$$\ln \frac{C_L}{C_o} + 2K_a(C_L - C_o) + \frac{K_a^2}{2}(C_L^2 - C_o^2) = -\frac{\varepsilon_c}{\varepsilon_g} K_r \eta \tau$$

or

$$\frac{1 - y_L^2}{2\gamma^2} + \frac{2(1 - y_L)}{\gamma} - \ln y_L = \frac{\varepsilon_c}{\varepsilon_g} K_r \eta \tau \qquad (14)$$

The value of y_L is not given explicitly by Equation 14 so a trial and error procedure is needed. The most efficient method is to specify a value for y_L, and then seek a set of values for γ and $K_r\eta$ referring to the same temperature.

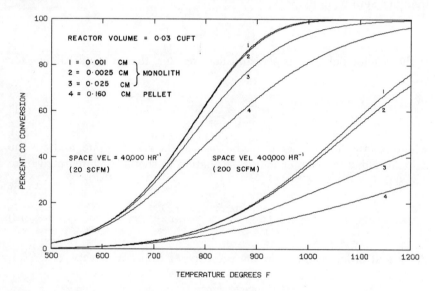

Figure 7. *Effect of catalyst thickness on CO conversion by base metal catalyst*

The computation is given for a monolith of 12-cm diameter and $L = 7.6$ cm. Let the square holes have a constant width of 0.25 cm. When the catalytic layer thickness l varies from 0.001 to 0.025 cm, the wall thickness varies accordingly to keep ϵ_g constant at 0.695. The calculation is extended to a packed bed of 0.32-cm pellets with uniform deposition of catalytic materials and $\epsilon_g = 0.4$. The spherical effectiveness factor is $\eta = 3(\phi \coth \phi - 1)/\phi^2$.

The first order kinetics for copper chromite catalyst is taken from Kuo *et al.* (6):

$$K_1 = K_1^\infty \exp(-E_r/RT)$$

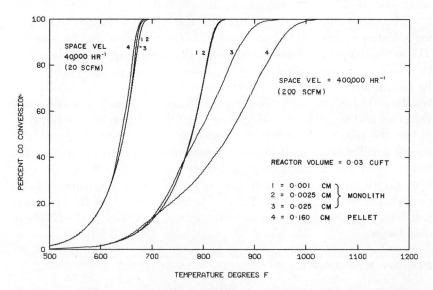

Figure 8. Effect of catalyst layer thickness on CO conversion for platinum kinetics

Inlet CO concentration, 1 mole %

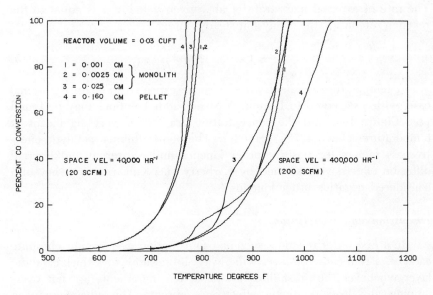

Figure 9. Effect of catalyst layer thickness on CO conversion for platinum kinetics

Inlet CO concentration, 4 mole %

where $K_1^\infty = 3.28 \times 10^8/\text{sec}$ and $E_r/R = 16{,}000\ °\text{R}$.

The calculated values for two different space velocities over base metal catalysts are plotted in Figure 7. They agree with the classical Thiele analysis, the thinnest layer being the best. Experimental data of this nature are often given in terms of the inlet temperature which can be several hundred degrees lower than bed temperatures in an adiabatic situation. Figures 8 and 9 present the findings for platinum at two space velocities and two inlet concentrations of CO. At 1% inlet concentration of CO, the kinetics is closer to first order kinetics; the thinnest layers still have a small advantage at higher temperatures and conversions, but the thicker layer has the advantage at lower temperatures. When the inlet concentration of CO is 4%, the optimum thickness is greater at 0.025 cm rather than 0.001 cm. The light-off temperature for the thicker layers is strikingly lower, corresponding to the maximum values of the effectiveness factor.

External Mass Transfer

When the reaction rates in the monolith channels are sufficiently high, a significant gradient will develop between the concentration at the channel center and that at the catalyst surfaces. This external mass transfer effect must be considered in addition to internal diffusion effects. The rate of external mass transfer at a given value of z is equal to the reaction rate inside the catalyst at steady state:

$$k_m a(C_g - C_s) = \frac{K_r \eta C_s}{(1 + K_a C_s)^2} \frac{\varepsilon_c}{\varepsilon_g} \qquad (15)$$

A temperature gradient would also be expected. For an isothermal case, with η set equal to 1, multiple steady-state solutions may be found (*see* Figure 10), and the concentration gradient is very significant at temperatures above 427°C (800°F). The non-isothermal catalytic effectiveness factors for positive order kinetics under external and internal diffusion effects were studied by Carberry and Kulkarni (*8*); they also considered negative order kinetics.

Discussion and Conclusions

In a first order reaction, such as CO oxidation over copper chromite catalyst, a thin catalytic layer will always be superior to a thick catalytic layer whenever the crystallites of the active ingredients are not overlapping and lead poisoning effects are absent. The advantage of a 0.001-cm layer over a 0.025-cm layer, as measured by CO conversion under identical temperature and inlet CO concentration, is not very great except when the space velocity is very large.

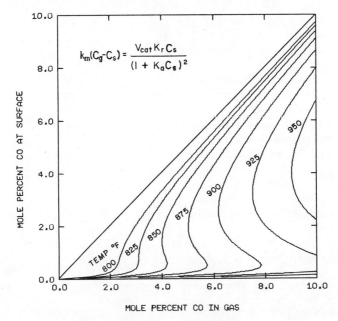

Figure 10. External mass transfer characteristics

In a negative first order reaction such as CO oxidation over platinum, a thick catalytic layer is often superior to a thin layer, especially when the space velocity of the gas is large and the inlet concentration of CO is high. However, these advantages are not very great when we measure results in terms of CO conversion efficiencies; it seldom amounts to more than a 28°C (50°F) shift in the temperature needed for 50% conversion. However, the effects of lead poisoning could increase this shift greatly, especially when lead deposition is concentrated in a zone of about 0.002 cm. Lead poisoning could also cause complete destruction of the egg shell catalyst but only a slight deterioration in the thick layer catalyst. It may be advantageous to have a sacrifice layer of alumina on the surface to adsorb lead and thereby protect the platinum in the interior.

The optimum distribution of catalytic material on the support depends on the kinetic order of the reaction catalyzed. When platinum is used for CO and hydrocarbon oxidation in automotive catalysis, the reverse of the normal wisdom is in order—platinum should be distributed toward the interior of the support layer to form a new type of egg yolk catalyst.

Nomenclature

a = external surface area per unit free volume of monolith channel, cm^{-1}
C = CO concentration, mole % at 1 atm total pressure

D_e = effective diffusivity, cm^2/sec
ΔH = heat of reaction, cal/mole CO
k_m = mass transfer coefficient, cm/sec
K_a = adsorption rate constant, 1/mole % CO
K_r = reaction rate constant, 1/sec
K_1 = reaction rate constant for first order reaction, 1/sec
l = pore length, cm
L = length of monolith or packed bed reactor, cm
R = gas constant
T = temperature, degrees Rankine
v = linear velocity of gas in reactor
x = length along pore, cm
$\bar{x}$ = dimensionless length along pore, x/l
y = dimensionless concentration, C/C_s
z = length along reactor, cm

Greek Letters

β = the Prater number
γ = $1/K_a C_s$ or $1/K_a C_o$
ε = volume fraction in reactor
η = effectiveness factor defined by Equation 5
λ = thermal conductivity of porous catalyst, cal/sec cm °K
ξ, ρ, σ = parameters in integration of Equation 10
τ = residence time, sec
ϕ = Thiele modulus defined by Equation 2

Subscripts

a = adsorption
c = porous catalyst
g = gas
L = outlet of reactor
m = mass
o = initial
p = provisional
r = reaction
s = surface
w = wall

Asymptotic Properties of Equations 3, 7, and 10

Equation 3

$$\frac{d^2 y}{d\bar{x}^2} = \phi^2 \gamma^2 \frac{y}{(\gamma + y)^2} \tag{16}$$

As $\gamma \to \infty$ (*i.e.*, $K_a \to 0$), Equation 16 reduces to the first order case:

$$\frac{d^2 y}{d\bar{x}^2} = \phi^2 y$$

Equation 7

$$\frac{dy}{dx} = \phi\, \gamma\, \sqrt{2\left[\frac{\gamma}{(\gamma+y)} - \frac{\gamma}{(\gamma+y_0)} + \ln(\gamma+y) - \ln(\gamma+y_0)\right]} \quad (17)$$

Let $\alpha = 1/\gamma$ and expand in series:

$$\begin{aligned}
\frac{dy}{dx} = \phi\, 1/\alpha\, \sqrt{2\,[}& 1 - \alpha y + (\alpha y)^2 - (\alpha y)^3 + (\alpha y)^4 - \cdots \\
& - 1 + \alpha y_0 - (\alpha y_0)^2 + (\alpha y_0)^3 - (\alpha y_0)^4 + \cdots \\
& + \alpha y - \frac{(\alpha y)^2}{2} + \frac{(\alpha y)^3}{3} - \frac{(\alpha y)^4}{4} + \cdots \\
& - \alpha y_0 + \frac{(\alpha y_0)^2}{2} - \frac{(\alpha y_0)^3}{3} + \frac{(\alpha y_0)^4}{4} - \cdots\,]
\end{aligned}$$

$$= \phi\, \sqrt{\frac{2}{\alpha^2}\left[\frac{(\alpha y)^2}{2} - \frac{(\alpha y_0)^2}{2} - \frac{2}{3}\alpha^3(y^3 - y_0^3) + \frac{3}{4}\alpha^4(y^4 - y_0^4)\cdots\right]}$$

$$= \phi\, \sqrt{y^2 - y_0^2 - \frac{4}{3}\alpha(y^3 - y_0^3) + \frac{6}{4}\alpha^2(y^4 - y_0^4)} \quad (18)$$

$\alpha \to 0$ reduces Equation 18 to the first order case:

$$\frac{dy}{d\bar{x}} = \phi\, \sqrt{y^2 - y_0^2} \quad (19)$$

Equation 10

$$I = \frac{1}{\phi\gamma}\int_{y_0}^{1}\frac{d\sigma}{\sqrt{2\left[\ln\left(\frac{\gamma+\sigma}{\gamma+y_0}\right) - \frac{\gamma(\sigma-y_0)}{(\gamma+\sigma)(\gamma+y_0)}\right]}} \quad (20)$$

From Equation 18,

$$I = \frac{1}{\varphi}\int_{y_0}^{1}\frac{d\sigma}{\sqrt{\sigma^2 - y_0^2 - \frac{4}{3\gamma}(\sigma^3 - y_0^3) + \frac{6}{4\gamma^2}(\sigma^4 - y_0^4) - \cdots}}$$

Letting $\zeta = (\sigma - y_0)$,

$$= \frac{1}{\phi}\int_{0}^{1-y_0}\frac{d\zeta}{\sqrt{\zeta}\underbrace{\sqrt{(\zeta+2y_0) - \frac{4}{3\gamma}[(\zeta+y_0)^2 + (\zeta+y_0)y_0 + y_0^2] + \cdots}}_{f(\zeta,y_0) > 0 \text{ for all } \zeta \text{ when } y_0 \neq 0}}$$

As $\zeta \to 0$,

$$I = \frac{1}{\phi f(0,y_0)}\int_{0}^{1-y_0}\frac{d\zeta}{\sqrt{\zeta}} = \frac{1}{2f(0,y_0)\,\phi}\sqrt{\zeta}\,\Big|_{0}^{1-y_0}$$

i.e., the integrand behaves like $1/\sqrt{\zeta}$ as $\zeta \to 0$, and it is not singular at $\zeta = 0$. However as $y_0 \to 0$, convergence of the numerical integration will be slow, and a modest degree of accuracy has to suffice. The integration is performed iteratively until sufficiently small step size is reached to ensure accuracy of the integral to four significant figures. For $y_0 \leq 0.0001$, the effectiveness curve is well represented by the asymptote equation.

Literature Cited

1. Voltz, S. E., Morgan, C. R., Liederman, D., Jacob, S. M., *Ind. Eng. Chem. Prod. Res. Develop.* (1973) **12**, 294–301.
2. Satterfield, C. N., "Mass Transfer in Heterogeneous Catalysis," Massachusetts Institute of Technology, Cambridge, 1970.
3. Roberts, G. W., Satterfield, C. N., *Ind. Eng. Chem. Fundam.* (1966) **5**, 317–325.
4. Petersen, E. E., "Chemical Reaction Analysis," Prentice-Hall, Englewood Cliffs, 1965.
5. Wei, J., *Chem. Eng. Sci.* (1965) **20**, 729–736.
6. Kuo, J. C. W., Morgan, C. R., Lassen, H. G., Soc. Automot. Eng. *Trans.* (1971) vol. 80, paper **710289**.
7. Weisz, P. B., Hicks, J. S., *Chem. Eng. Sci.* (1962) **17**, 265.
8. Carberry, J. J., Kulkarni, A. A., *J. Catal.* (1973) **31**, 41–50.

RECEIVED May 28, 1974. Research supported by National Science Foundation grant GK-38189.

11

Oxidative Automotive Emission Control Catalysts—Selected Factors Affecting Catalyst Activity

LOUIS C. DOELP, DAVID W. KOESTER, and MAURICE M. MITCHELL, JR.

Air Products and Chemicals, Inc., Houdry Division, Linwood, Pa.

Certain factors are analyzed to determine their effects on automotive catalyst activity. At operating gas velocities, spherical catalysts were more active than monolithic catalysts at comparable catalyst volumes and metals loadings. Palladium was the most active catalyst metal. Platinum in a mixed platinum–palladium catalyst stabilizes against the effects of lead poisoning. An optimum activity particulate catalyst would contain about 0.05 wt % total metals on a gamma-alumina base with a platinum content of 0.03–0.04 wt % and a palladium content of 0.01–0.02 wt %. A somewhat thick shell of metals located near the outer surface of the particle provides better catalyst activity than a shell type distribution of metals.

In the development of oxidative automotive emission control catalysts for use in the 1975 model year, certain requirements were recognized from the beginning. The catalyst had to be physically rugged and capable of withstanding both the mechanical and thermal abuse to which it would be subjected in an automobile driven by average drivers on real roads. The catalyst had to exhibit high levels of activity so that the catalytic units would be of reasonable size. The catalyst had to be stable for at least 50,000 miles and capable of withstanding chemical abuse from the exhausts of the various fuels to which it would be subjected.

Certain physical and chemical factors emerged as important in meeting these requirements. These physical properties included crushing strength, attrition resistance, shrinkage upon thermal exposure, bulk

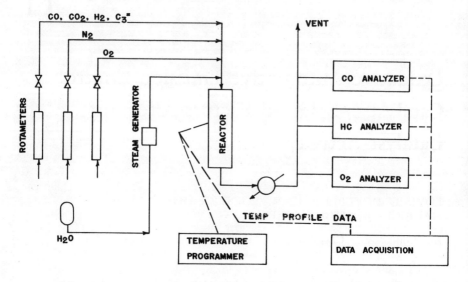

Figure 1. Schematic of automotive emissions catalyst test unit

density, and, of course, surface area and pore volume; these properties are characteristic of the support material and are not affected, except minimally, by the active precious metal components added to it.

In this paper we report on some of the factors that affect catalyst performance in the presence of lead and sulfur both before and after high temperature aging. These are total metals loading, catalyst compo-

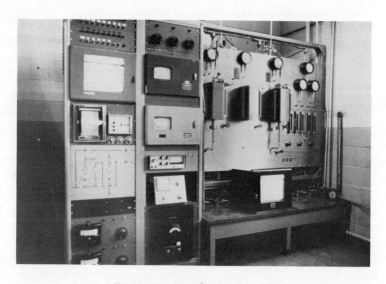

Figure 2. Bench test assembly

sition, and metals distribution between catalyst particles as well as within the catalyst particle. An interesting aspect of the difference between particulate and monolithic catalysts is also discussed.

Experimental

Catalyst activity was usually measured in a bench test assembly (Figure 1). The reactor included a preheat section containing tabular alumina just above (upstream from) the 30 cm^3 of catalyst in the center of the reactor. Water was pumped by a minipump (Milton-Roy) to the steam generator. From a three-temperature profile around the catalyst bed, it was determined that the midpoint data were most useful and reliable. The analytical equipment consisted of an infrared device (Mine Safety Appliances) for carbon monoxide, a flame ionization detector (Beckman) for hydrocarbons, and a paramagnetic oxygen analyzer (Beckman). The entire assembly except for Telex printer and computer is pictured in Figure 2.

All data were collected through the data acquisition system, digitized, and stored on paper tape. These tapes were batch-fed into a computer which analyzed the data and fed back the required conversion *versus* temperature profiles. The data were also displayed on strip chart recorders.

Table I. Gas Composition in Auto Emissions Catalyst Test Unit

Component	Content, vol %
CO_2	9.0
CO	1.35
H_2	0.45
$C_3^=$	0.0225
O_2	2.7
Steam	10.0
N_2	balance

Composition of the gas fed to the preheat section of the reactor (Table I) was selected by averaging data from numerous sources on the composition of automobile exhaust gases. Propylene simulated all hydrocarbons in the exhaust gas. For some experiments, the catalysts were aged in a 260-in.^{3} converter attached to the exhaust pipe from a 350-in.^{3} V8 engine under a water brake dynamometer load equivalent to 50 mph.

Monoliths vs. Spheres

In order to determine the relative efficiency of monoliths and spheres given a fixed converter volume constraint, a standard Corning cordierite monolith (1.52-mm square hole, 0.25-mm wall) and some γ-alumina spheres 2.5–4.0 mm in diameter were coated in the same way with a washcoat which gave 6 wt % γ-alumina after firing. Spheres and mono-

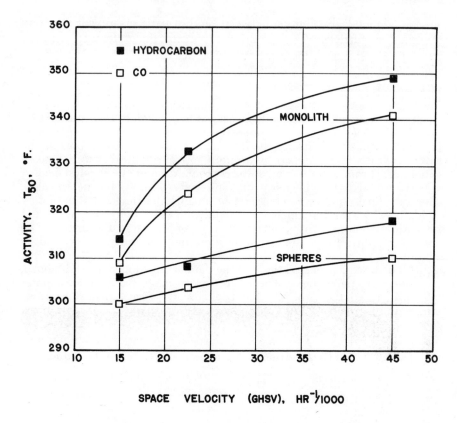

Figure 3. Activities of spherical and monolithic catalysts with equal catalyst volumes and metals loadings

lith were impregnated with palladium at a loading of 0.255 troy oz/ft³ (0.280 g/l). In the alumina sphere catalyst, this amounted to about 0.038 wt % palladium.

The bench test activity of these two systems is plotted *versus* space velocity in Figure 3. Activity is measured by the T_{50}, that temperature at which 50% conversion is achieved. These data were all collected during continuous cooling from an initial temperature of 593°C (1100°F). At the higher velocities, the spheres were more active per given volume of total catalyst. The reproducibility of the bench test is indicated by standard deviations of about 13°F for the carbon monoxide activity and about 11°F for the hydrocarbon activity. It can be argued, therefore, that at the lower velocities [15,000 gas hourly space velocity (GHSV)], which correspond to little more than idling velocities (12–14 mph), there was essentially no difference between monolith and sphere. But certainly at the practical velocities of an automobile in motion (GHSV of 22,000–50,000), the spheres had a distinct advantage, at least on a volume efficiency basis.

Optimum Metals Loading and Composition

The T_{50} activity of impregnated γ-alumina spheres is plotted vs. palladium content in Figure 4. Pure palladium catalysts as well as platinum and palladium co-impregnated catalysts were tested. When palladium concentration was increased above 0.03 wt %, there was no further increase in CO oxidation activity. Furthermore, the activity of catalysts containing both platinum and palladium, with the platinum:palladium weight ratio varying tenfold, was virtually unaffected by the presence of platinum.

The activity for carbon monoxide oxidation was plotted versus varying platinum and palladium concentration at a fixed total metals concen-

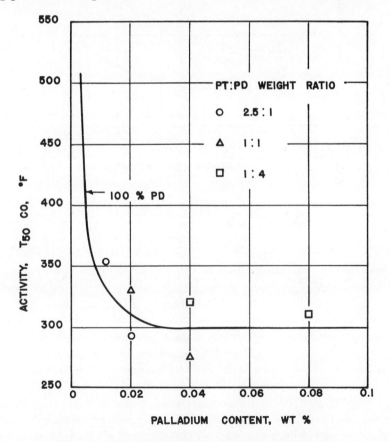

Figure 4. CO oxidation activity vs. palladium content

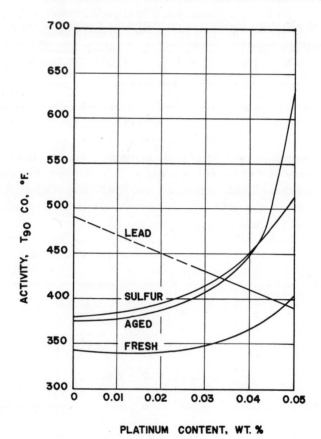

Figure 5. Effects of lead, sulfur, and aging on CO activity

Platinum + palladium = constant 0.05 wt %

tration of 0.05 wt % (Figure 5). Because the T_{50} data tend to converge regardless of the condition of the fresh catalyst, the data plotted are for 90% conversion (T_{90}). Both pure platinum and pure palladium at the same total metals loading had disadvantages. Palladium was extremely sensitive to lead poisoning whereas platinum was sensitive to sulfur. Platinum was also sensitive to thermal aging, probably through a sintering mechanism. Aging was achieved by exposure to 982°C for 24 hrs in convective air. Up to very high platinum contents, the effects of sulfur and of thermal aging were almost indistinguishable.

A similar plot of hydrocarbon oxidation activity is presented in Figure 6. The activity–composition relations reflect a more extreme sensitivity of pure palladium toward lead in hydrocarbon oxidation than was observed in CO oxidation (cf. Figure 5). The difference between palladium and platinum resistance to sulfur and thermal aging was more distinct in hydrocarbon oxidation.

The concentrations of lead and sulfur on the catalysts characterized in Figures 5 and 6 were 0.2 and 2–4 wt % respectively. These levels correspond to about 3000–5000 miles with 1975 model year automobile fuel (0.03–0.05 g Pb/gal). The lead was impregnated onto the fresh catalyst using lead nitrate in order to obtain more nearly solely chemical poisoning effects rather than the pore mouth blocking effects of lead oxide from exhaust gas. After impregnation, the leaded catalyst was heat treated 4.5 hrs at 482°C. For the sulfur, the catalyst was treated at 482°C in an air stream containing 5% steam and 1% SO_2.

It appears from Figures 5 and 6 that there may be a platinum:palladium ratio at constant metals concentration which gives optimum

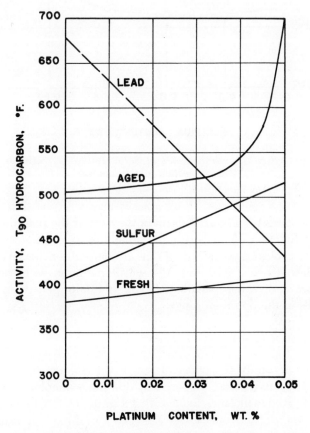

Figure 6. Effects of lead, sulfur, and aging on hydrocarbon activity

Platinum + palladium = constant 0.05 wt %

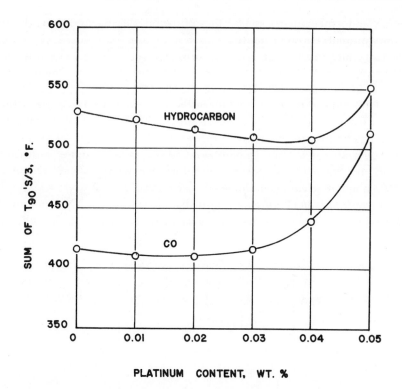

Figure 7. Sum of hydrocarbon and CO activities vs. platinum content at constant metals loading

Platinum + palladium = constant 0.05 wt %

activity and stability. Figure 7 displays the sum of the poison and aging effects as a function of this platinum:palladium ratio, again at constant total metals concentration. The optimum composition for hydrocarbon oxidation lies between 0.03 and 0.04 wt % platinum with the balance (to 0.05 wt %) being palladium. The platinum:palladium ratio of 2.5: 1.0 falls in the optimum range.

Metals Distribution between Spheres

In order to study the effect of metals distribution on activity and stability, test catalysts were prepared by diluting γ-alumina sphere-based catalysts of fixed concentration (a Pt:Pd ratio of 2.5:1.0) with bare spheres in varying amounts. The T_{50} data for these test catalysts for CO oxidation are plotted in Figure 8. Each curve represents the data at a constant total metals concentration when calculated over the entire catalyst–bare sphere combination. The curves demonstrate the effect of sphere-to-sphere maldistribution, and this effect, though real, was not very great for very minor variations in distribution. Aged catalyst was more affected by maldistribution than fresh catalyst. The test catalysts had low metals loadings of 0.025 and 0.0125 wt %. On the right side of the figure (at 100%) are plotted the 0.05 wt %, uniform distribution data to give some indication of the relative effect of concentration.

A similar plot depicting the effects on hydrocarbon oxidation efficiency of dilution of the catalyst concentration is presented in Figure 9. The conclusion is similar—minor variations at close to uniform distribution are not critical, and the effect is greater for aged catalyst. The effect of maldistribution on aged catalyst was greater for hydrocarbon than for carbon monoxide oxidation (*cf*. Figure 8).

Metals Distribution within the Catalyst Sphere

X-ray spectra (*1*) were made of the surfaces of two similar catalysts (Figure 10). Both catalysts contained 0.05 wt % total metals with a platinum:palladium ratio of 2.5:1.0. They were made in the same way

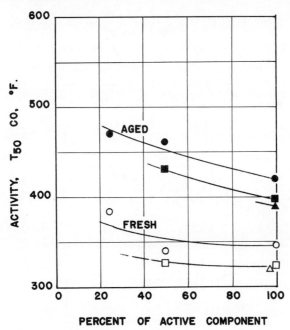

Figure 8. Effect of catalyst bed dilution on CO activity.

Platinum:palladium = 5:2. Total metals loading: ○, 0.0125 wt %; □, 0.025 wt %; and △, 0.05 wt %. Open shapes, fresh catalysts; and solid shapes, aged catalysts.

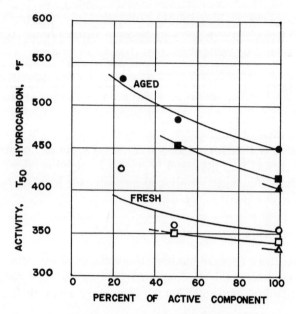

Figure 9. Effect of catalyst bed dilution on hydrocarbon activity

Conditions and key as in Figure 8

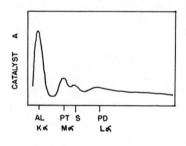

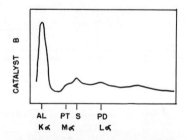

Figure 10. Expanded energy dispersive x-ray analysis of catalyst surface

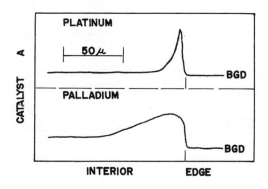

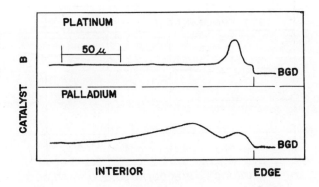

Figure 11. Relative Pt and Pd concentration profiles determined by electron probe studies of polished cross sections

For activities of catalysts, see Table II

except that there was some variation in the precursor metal complexes used in impregnation. The surface concentrations of platinum and palladium were higher in catalyst A than in catalyst B.

Figure 11 is a plot of some radial distributions (1) of the platinum and palladium in these same two catalysts. The metals concentrations

Table II. Bench Test Activity of Catalysts A and B

Catalyst	Condition	Activity, T_{50} °F	
		CO	HC
A	fresh	320	326
	aged	394	404
B	fresh	299	306
	aged	365	373

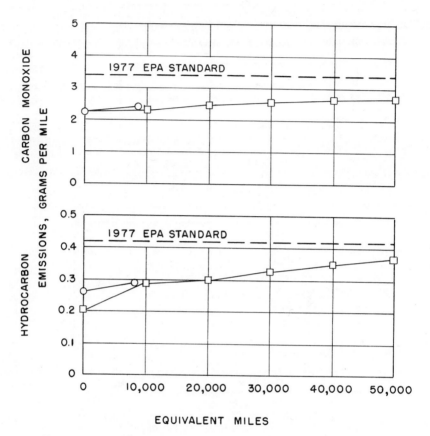

Figure 12. Hydrocarbon and CO emissions vs. catalyst age
○, *catalyst A and* □, *catalyst B*

are not proportional to the areas under the curves because the spectral sensitivities of platinum and palladium differ. The indications from Figure 10 are confirmed: catalyst A has a higher proportion of its platinum and palladium near the surface than does catalyst B. The activities of fresh and aged catalysts A and B are tabulated in Table II. The bench test activities of both fresh and aged catalysts for both CO and hydrocarbons were superior for catalyst B.

The emissions with these two catalysts were plotted *vs.* hours aged on an engine dynamometer expressed as the equivalent number of miles at a steady rate of 50 mph (Figure 12). Included in the plot are the original 1975 (now 1977) Environmental Protection Agency (EPA) requirements. There was little difference in initial activity for carbon monoxide that resulted from different platinum and palladium distributions, but there was a distinctly different fresh activity for hydrocarbons which was consistent with the bench data in Table II.

Some synthetic means of aging catalysts have been presented in this paper, and their effects were described in terms of bench test activity measurements. It would be of interest to have some perspective on how these activities relate to actual aging data. Although it is recognized that T_{50} and T_{90} bench test data may not uniquely correlate with EPA test procedure (1975 CVS) data, the bench test data in this paper were confirmed by the emission data in Figure 12. Furthermore, samples of catalyst B were aged for different periods on an engine dynamometer. Each sample was then evaluated in the bench test unit. Bench test activities are presented in Figure 13 as a function of equivalent miles of exposure at a steady rate of 50 mph. The continuous, smooth curve expresses actual aged catalyst performance in terms of bench test activity.

Summary and Conclusions

Some of the factors affecting the activity and durability of automotive emissions control catalysts were studied. It was found that at

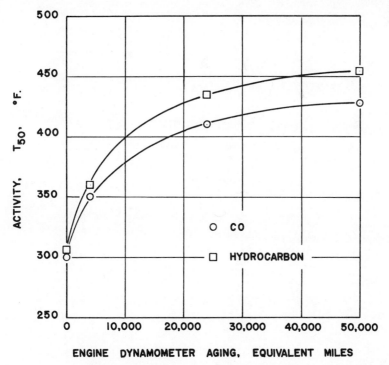

Figure 13. Correlation of bench activity data with engine dynamometer aging of catalyst B

the same volume of catalyst and same metals loading, spherical catalysts are more active than monolithic catalysts. The major role of platinum in platinum–palladium catalyst is to decrease activity loss resulting from lead poisoning. From a consideration of several modes of catalyst deactivation, it appears that optimum catalyst performance is achieved with a platinum:palladium ratio of 1.5:1–4:1. There is some evidence that a finite, thick, shell deposition of metals in the spherical particle is better than a very thin surface shell.

Acknowledgment

The authors acknowledge the experimental assistance of their colleagues. Especially worthy of mention are W. Alexander, G. D. Cooper, S. M. Stetina, L. L. Upson, and K. D. West.

Literature Cited

1. Ficca, Jr., J. F., Micron, Inc., Wilmington, Del., private communication.

RECEIVED September 19, 1974.

12

Thermally Stable Carriers

R. GAUGUIN, M. GRAULIER, and D. PAPEE

Rhône-Poulenc, 21, rue Jean-Goujon, 75008 Paris, France

The textural and mechanical properties of carriers for auto emission catalysts should remain stable after exposure to high temperatures. Most high surface aluminas undergo nonhomogeneous crystallite growth and loss of porosity and mechanical strength when heated to 1000°C. Conversely, a single-phased pure alumina prepared from finely crystallized α-monohydrate which first transforms into γ-alumina, has the required stability; it still consists of a homogeneous δ-alumina phase at 1000°C. Ordinary aluminas cannot be sufficiently stabilized by incorporation of foreign oxides. However, some rare earth oxides, when added to γ-alumina, extend its stability to 1250°C by forming a θ-alumina phase, the specific surface of which still exceeds 25 m^2/g. Such compounds offer new possibilities for catalysts submitted to extreme temperatures.

Numerous catalyst carriers have been developed and manufactured on an industrial scale for use in fixed bed reactors. The carriers are characterized mainly by the nature of the product, the texture, the mechanical strength, and the stability. It is the nature of the product, typically an oxide, that confers surface properties which may be acidic, basic, or neutral. Depending on the application, Al_2O_3, SiO_2, TiO_2, ZnO_2, SiO_2–MgO, SiO_2–Al_2O_3, Al_2O_3–MgO, or activated carbon is used. The texture varies widely. For example, the surface area of activated carbon exceeds 1000 m^2/g whereas that of silica reaches that level; aluminas may have a surface area varying from a fraction of a square meter to 450 m^2/g and a pore volume of 0.15 to 1.0 cm^3/g. Mechanical strength, which is measured by crushing and attrition tests, must be appropriate for bulk-loading and use in a catalytic reactor. All these properties must be stable under working conditions at generally moderate temperatures of 300°–500°C with a maximum of 600°C. These usual requirements are generally met without difficulty.

Table I. Typical

Carrier	Surface Area, m^2/g		Crushing Strength,[b] kg		Attrition[c], %	
	Fresh	Calcined	Fresh	Calcined	Fresh	Calcined
A_2	250	60	10	3	1	4
A_3	250	70	9	2.5	3	8
B_2	250	85	10	6	0.5	1
B_3	250	80	8.5	4	3.5	4
C_1	90	50	10	10	0.2	0.5
C_2	90	50	8	8	0.5	1
D	250	80	9	9	0.5	0.5
E_1	10	10	10	10	2	2
E_2	10	10	10	10	3	3

[a] Fresh alumina or after 24-hr calcination at 1000°C.
[b] Crushing strength: measured on individual balls.
[c] Attrition: % of fines formed after vibration in a steel container.

The problems encountered with catalysts for auto emission purification are of the same type but of a totally different magnitude. Superior mechanical properties and, specifically, exceptional resistance to attrition are required. Furthermore, these properties should be unaffected by exposure to 1000°C and even 1100°C. Furthermore, a surface area of 50–80 m^2/g and porosity of 0.6–0.8 cm^3/g must remain after heating. Naturally the carrier must also withstand thermal shocks, offer the least possible diffusional limitation to reagents, and react as little as possible —or at least without unfavorable results—with the catalytically active oxides and metals deposited on its surface.

Several research approaches and industrial developments have been attempted in order to achieve these objectives. We shall consider the chemical, textural, and mechanical properties and the thermal stability of the main alumina carriers manufactured by Rhone-Poulenc as well as the effects of adding various oxides on these properties and on the performance of oxide-based catalysts.

Alumina Carriers

The mechanical properties of some carriers (balls, 2.4–4 mm in diameter) together with their chemical analysis and texture are listed in Table I. The mechanical properties of fresh carriers (*i.e.* those not exposed to 1000°C) were good; surface area was 5–10, 80–100, and 200–350 m^2/g, and pore volume was close to 0.5–0.6 cm^3/g. After calcination at 1000°C, notable differences in the mechanical properties were apparent. The target of 5-kg crushing strength and less than 1% attrition, con-

Properties of Carriers[a]

Total Pore Volume[d], cm³/g	Macropore Volume[d,e], cm³/g	Na_2O, %	SiO_2, %	Structure X-ray Diffraction
0.55	0.10	0.1	0.1	mixture of transition aluminas
0.65	0.15	0.1	0.1	
0.55	0.10	0.2	2	mixture of transition aluminas
0.65	0.15	0.2	2	
0.65	0.05	0.1	0.1	gamma
0.70	0.10	0.1	0.1	gamma
0.80	0.30	0.1	0.1	gamma
0.40	—	0.1	0.1	alpha
0.30	—	0.2	1.2	alpha

[d] For fresh alumina.
[e] $\phi > 1000$ A.

sidered sufficient on the basis of industrial experience, was attained only with carriers B2, C1, C2, and D. If a pore volume of at least 0.6 cm³/g is required, only carriers C1, C2, and D would still qualify. These three have the additional advantage of purity; as we shall see later, in many cases SiO_2 and Na_2O should be avoided. What causes such variation in the mechanical properties of one carrier but not of another?

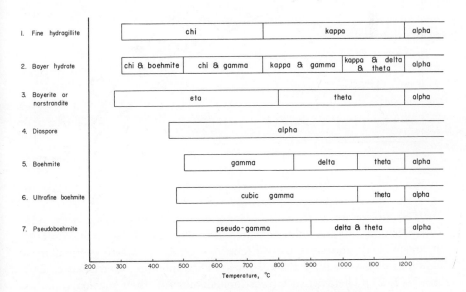

Figure 1. Transformation sequences of aluminas

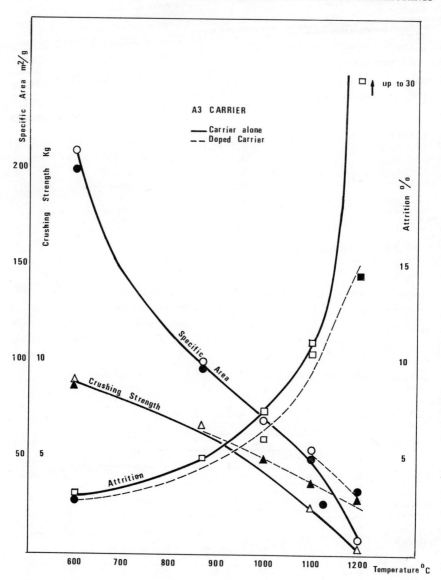

Figure 2. Characteristics of A3 carrier

Chemically Pure Aluminas. Depending on the precursor hydroxide, the transition aluminas obtained by calcination at 450°–1100°C differed in morphology and structure (Figure 1) (*1, 2, 3, 4, 5*). In fact, most aluminas consist of a mixture of phases corresponding to sequences 2, 3, and 7 (*1, 2*). This was so for carriers A2 and A3: their thermal stability was poor. On the other hand, aluminas C1 and C2 (sequences 5) were stable up to 1050°–1100°C (*cf.* solid lines in Figures 2 and 3); a number of reasons can be proposed to explain this remarkable stability.

First, note that any calcination causes the crystallites to grow considerably. An active alumina may have an original surface area of 350 m^2/g; after exposure to 1000°C the surface area decreases to 60 m^2/g.

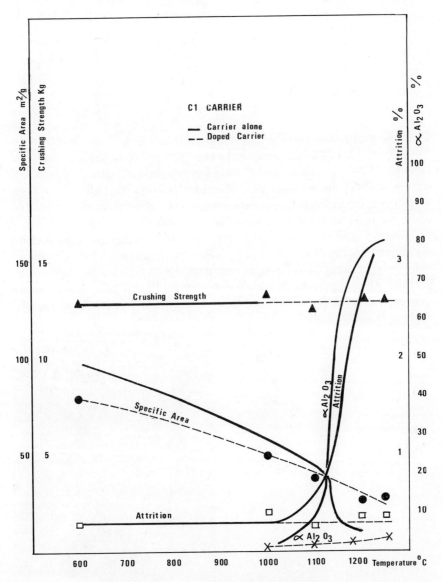

Figure 3. *Characteristics of C1 carrier*

Table II. Effect of Crystallite Size

Sample	Structure of Hydroxide	Surface of Boehmite, m^2/g	Crystallite Homogeneity	Attrition at 600°C, %	Attrition at 1000°C, %
1	hydrargillite + bayerite + pseudo-boehmite	—	apparently good but heterogeneous structure	1–4	6–8
2	boehmite	10	poor	2	5
3		30	poor	1.5	4
4		30	good	1	2
5		70	good	<0.5	<0.5
6		70	very good	<0.5	<0.5
7		200	very good	<0.5	<0.5

Thus the crystallites undergo growth by a factor of 6: from 50–100 A to 300–600 A. At 1200°C, after α-alumina has formed, particle size approaches 5000 A. This drastic modification brings with it significant strains; it alters the microscopic edifice and destroys the balls. Therefore, it is important to keep the variation in surface area as small as possible when the product is calcined from 500° to 1000°C.

Second, and even more important, this sintering is accompanied by crystalline modifications (Figure 1). The transformation of an aluminum hydroxide to corundum through transition aluminas causes a crystalline disorder, the amplitude of which depends on the specific structural sequence. Some of the aluminas consist of mixtures that undergo parallel

Figure 4. Sample 1 in Table II calcined at 870°C

Magnification, × 50,000

Figure 5. Sample 1 in Table II calcined at 1000°C
θ- and α-aluminas; magnification, × 50,000

transformations. Maintaining good mechanical cohesion in a porous alumina carrier with a large surface area over a broad temperature range certainly implies regular growth of the microcrystals of transition alumina, at least according to certain crystallographic directions. The exact lattice structure of transition aluminas is not fully understood. Yet it is remarkable that the sequence that leads from boehmite to the gamma phase and then from the gamma to the delta phase between 800°C and 1000°–1050°C produces aluminas with stable mechanical properties in that temperature range. Delta appears as a superstructure of three elementary cells of tetragonal gamma spinel transition alumina which itself is relatively well ordered in the anion cubic lattice but disordered especially in the distribution of the cations (6). Consequently, the formation of δ-alumina is rendered fairly easy by calcining boehmite at temperatures above 800°C. This transformation takes place without basic modification of the structure of the alumina, consequently without altering the mechanical properties. On the contrary all the aluminas for which a theta phase appears around 800°C lose their mechanical strength.

Third, the homogeneity of the starting hydroxide is a critical factor, even when the crystallite structure is boehmite. For example, with various boehmites, the products were more attrition resistant and more stable as the fineness and homogeneity of the crystallites of the precursor hydroxide increased (Table II).

These three observations—amplitude of variation in size of the crystallites during calcination, phase changes, and homogeneity—help to explain the differences in the stability of various aluminas. A comparison

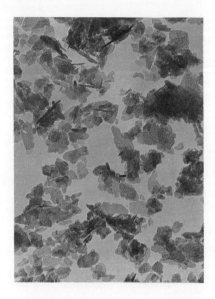

Figure 6. Sample 5 in Table II calcined at 600°C

Magnification, × 50,000

of Figures 4 and 5 (Sample 1 in Table II) with Figures 6 and 7 (Sample 5 in Table II) is revealing. The heterogeneity in crystal size after calcination of the Sample-1 product at 870°C was startling. With Sample 5, no significant change was observed between 600° and 1000°C; crystallite growth and heterogeneity appeared only at 1200°C.

Table III. Effect

Calcination at 870°C for 24 hrs

Carrier	Additive cmpd	wt %	Surface Area, m^2/g	Crushing Strength, kg	Attrition, %	Structure
A3	—	—	97	6.5	7	theta + delta
	CuO	7	85	3	8	delta + 5% alpha
	MoO_3	10	—	< 0.5	—	alpha
	V_2O_5	7	—	—	—	—
	Cr_2O_3	5.2	96	6	9.5	theta type
	Na_2O	4.8	88	5	—	—
	TiO_2	2	—	—	—	—
	SiO_2	4.8	120	7	3.4	—
	CdO	8.6	70	5	—	cubic, eta type
	ZnO	6.4	132	7	3	cubic
	CeO_2	11	100	6	7	cubic + CeO_2
	MgO	2.4	135	7	6	cubic
C1	—	—	63	10	0.3	delta
	CuO	7	50	10	0.5	delta
	CeO_2	11	50	10	—	—
	MgO	2.4	70	10	0.3	—

Figure 7. Sample 5 in Table II calcined at 1000°C
δ-alumina; magnification × 50,000

Remedies for Instability. One possible remedy for improving the stability at 1000°C of some carriers is to add oxides (introduced by impregnation as soluble components) which react with the alumina to form stable compounds.

of Additives

Calcination at 1000°C for 24 hrs

Surface Area, m^2/g	Crushing Strength, kg	Attrition, %	Structure
70	2.5	8	theta + alpha + delta
7	1	40	70% alpha + $CuAl_2O_4$
0			
1.5	1	2	theta + alpha + V_2O_5
58	3	15	theta type + 5% alpha
67	4	8	gamma + beta
60	5	17	—
80	4	4	delta + theta
37	3	14	theta + delta + alpha
53	3	14	theta + alpha
66	4	9	theta + CeO_2
97	4	8	delta + $MgAl_2O_4$
50	10	0.5	delta + theta
10	6	1.5	70% alpha + $CuAl_2O_4$
41	10	0.3	—
57	15	0.2	$MgAl_2O_4$ + delta + alpha

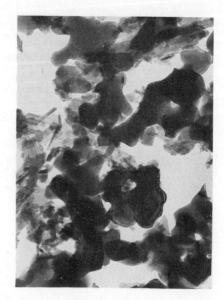

Figure 8. Sample 5 in Table II calcined at 1200°C

θ- and α-aluminas; magnification × 50,000

In this study, we pursued a double objective: to improve the mechanical and textural properties of alumina calcined at 1000°C and to investigate the possible consequences at this temperature of interactions of alumina with certain oxides normally used in the manufacture of oxidation catalysts. Included in the latter category are CuO, MoO_3, V_2O_5, and Cr_2O_3. The carriers tested (Table III) were A3 which is particularly sensitive to calcination above 850°C and C1 which is already stable.

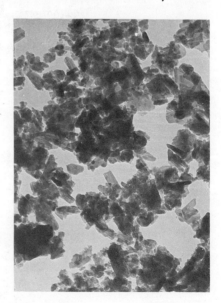

Figure 9. Carrier C1 + rare earth oxide calcined at 1200°C

θ-alumina + hexagonal aluminate; magnification, × 50,000

Oxides of copper, molybdenum, and vanadium had a disastrous effect on the carriers. Surface area dropped to less than 10 m^2/g, and crushing strength was much less than that of the original sample. Degradation by molybdenum oxide is well known: at 600°C aluminum molybdate is formed and decomposes at about 800°C into α-Al$_2$O$_3$ and MoO$_3$ which sublimates.

With copper oxide, a mixed phase was also obtained; a poorly crystallized copper aluminate containing a large excess of alumina. This structure was stable at 900°C, but between 900° and 1000°C stoichiometric copper aluminate was formed and excess alumina was rejected from the structure as α-Al$_2$O$_3$. A2 and A3 carriers could not withstand this transformation; types C1 and D were less degraded.

- It is known that silica retards the appearance of α-Al$_2$O$_3$, and carriers of the B series utilize this property. SiO$_2$ may be a poison for certain catalysts (see below).

- Cr$_2$O$_3$, CdO, and Na$_2$O appreciably reduced resistance to attrition.

- MgO and CeO$_2$, and ZnO to a lesser extent, act as stabilizers. The effect of ZnO was appreciable only below 900°C. The effect of MgO was stronger and increased with the heterogeneity of the carrier; the surface area left after calcination at 1000°C increased from 70 to 97 m^2/g for carrier A3 but only from 50 to 57 m^2/g and from 85 to 90 m^2/g for carriers C1 and D respectively.

Interpretation. It is assumed that the incomplete occupancy of tetrahedral sites could be responsible for the metastability of transition aluminas. It is possible to retard the diffusion of Al^{3+} cations which are responsible for the structure transformations by introducing bivalent ions such as those of Cd, Ca, Zn, Mg, Ni, and Cu into the tetrahedral position in the spinel cubic lattice.

If without additives, a carrier such as A3 consists of a mixture of δ-, θ-, and α-aluminas between 900° and 1000°C. The presence of the metal oxides, introduced by impregnation, effectively maintains a cubic type structure at a calcining temperature of 900°–1000°C. With alumina, these oxides form spinel-structured compounds which are more or less well crystallized. Because of the insertion of alumina, the lattice parameter of these compounds is expanded with respect to that of the stoichiometric spinel. The properties of the carrier are thus maintained up to a temperature which depends on the considered mixed oxide. At this temperature—about 1000°C with magnesium aluminate and 900°C with zinc and copper aluminates—the stoichiometric spinel recrystallizes while α-alumina is rejected (6). The carrier then suddenly loses its mechanical and structural properties. None of the mentioned additives could improve the stability of the C1 carrier above 1000°C.

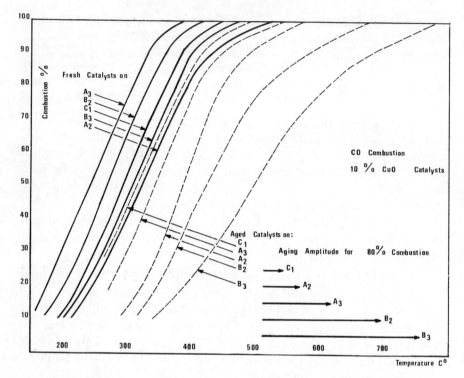

Figure 10. Effect of catalyst aging on CO combustion

A positive effect was obtained with some rare earth oxides. They behaved differently and favored the formation of a theta-type structure stable at high temperature. Consequently, the transformation into α-alumina was postponed, and a surface of approximately 30 m^2/g was kept at 1200°C (Figures 8 and 9). For type A carriers, the mechanical properties were little improved, but with a C1 alumina a remarkable attrition resistance was maintained (*cf.* dotted lines in Figures 2 and 3).

In summary, the stability of aluminas is largely dependent on phase transformations. Carriers of the A type exhibit progressive degradation of their mechanical properties in relation to the appearance of θ-alumina above 800°C. Some additives provide a slight improvement; the progressive degradation is not avoided, only reduced. Carriers of the C type behave differently. The sequence gamma → delta provides great stability of the mechanical properties for temperatures about 1000°C. If temperatures above 1050°C are required, the fast, almost simultaneous, transformation delta → (theta) → alpha must be avoided. This is achieved when a stable theta phase is formed by adding some rare earth oxides; the mechanical properties are not modified, but their stability is extended up to 1200°C.

Reactions of Activated Oxides with Carriers—Catalytic Properties

The addition of a few wt % oxides to an alumina carrier has considerable effect and appreciably modifies the carrier's mechanical and textural properties. The finished catalyst, if it consists of active oxides, will have to withstand successfully any reactions which might take place between alumina and the oxides during the high temperature exposure. The instability of Cr_2O_3, and especially of CoO, is typical. By contrast, CuO maintains a good level of activity.

The stability of CuO is dependent on the quality of the carrier. It was established that the specific surface area of the carrier has little effect. Copper oxide catalysts prepared on A3 (250 m^2/g), C1 (90 m^2/g), and E1 (10 m^2/g) had similar activity after calcination at 1000°C. On

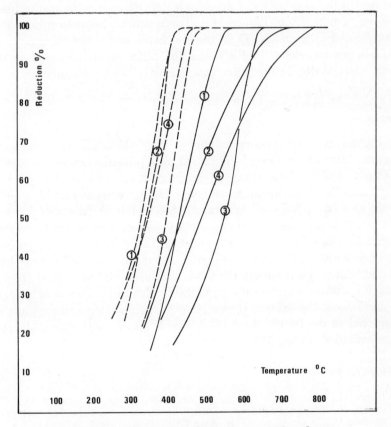

Figure 11. Effect of catalyst aging on NO reduction.

Catalysts with 10% Cr_2O_3, Fe_2O_3, and NiO on carriers 1, E1; 2, E2; 3, C1; and 4, B2 with surface adjusted to 60 m^2/g. Exhaust gas: 0.2% NO, 2% CO, 1% H_2, and 3% H_2O in N_2; VSHV, 20,000. - - - - -, fresh catalyst; and ———, catalyst aged 24 hrs at 1000°C.

the other hand, the presence of soda or silica was very detrimental (*cf.* A2 and A3, B2 and B3 in Figure 10). Experiments were conducted to confirm this effect. The addition of 0.5% Na_2O (caustic soda impregnation) or 1–3% SiO_2 (ethyl silicate or sol impregnation) to carrier C1 caused a drop in the performance of the catalyst treated at 1000°C and degraded it to the level of type B carriers.

With more complicated formulas involving *e.g.* Cr_2O_3, NiO, and Fe_2O_3, soda, silica or even an excessively large specific surface area may be a disadvantage. With such catalysts used for the reduction of nitrogen oxides, stability was again dependent on the carrier. The effect of SiO_2 and Na_2O alone was reflected in a 120°C differential between the temperatures required to achieve 95% conversion on catalysts calcined at 1000°C (*cf.* E1 and E2 in Figure 11). The effect of the alumina surface area was of the same order of magnitude (*cf.* B2 after surface adjustment to 60 m^2/g with E2 in Figure 11). The reactions between nickel ferrichromite and SiO_2 and Na_2O on the one hand, and transition alumina—whether gamma, delta, or theta—on the other hand help explain this effect. Specifically, in the latter case $NiAl_2O_4$ can be identified:

$$NiFe_{2-x}Cr_xO_4 + Al_2O_3 \longrightarrow NiAl_2O_4 + Fe_{2-x}Cr_xO_3$$

Conclusions

Carriers with satisfactory stability at 1000°–1050°C and a sufficiently developed surface area may be obtained from pure alumina of the gamma structure. Among the carriers investigated, type C offers the optimum compromise for auto emission catalysts. Lighter carriers such as type D should also be of interest. Whether noble metals or base metal oxides are used, it is necessary to choose a carrier which can be easily impregnated and which favors the mass and heat transfers in the catalyst bed. With noble metals, the choice of carrier determines by itself the mechanical and textural properties of the finished catalyst. If the catalyst consists of oxides, optimum chemical compatibility should be obtained between the carrier and the oxides. If operation at higher temperatures must be considered in the future, a modified type C carrier will offer a stability range extended up to 1200°C.

Literature Cited

1. Thibon, H., Charrier, J., Tertian, R., *Bull. Soc. Chim. Fr.* (1951) 384.
2. Tertian, R., Papee, D., *J. Chim. Phys. Physicochim. Biol.* (1958) 341.
3. Lippens, B. C., De Boer, S. H., *Acta Crystallogr.* (1964) **17**, 1312.
4. Alcoa, "Alumina Properties," *Tech. Paper* (1972) **19**.
5. Krichner, H., Torkar, K., Hornisch, P., *Monatsh. Chem.* (1968) **99**, 1733–41.
6. Lejus, A.-M., Thèse de Doctorat, Institution, 1964.

RECEIVED May 28, 1974.

/ 13

Spinel Solid Solution Catalysts for Automotive Applications

C. A. LEECH, III[1] and L. E. CAMPBELL[2]

Research and Development Laboratories, Corning Glass Works, Corning, N. Y.

> Enhancement of initial oxidation activity and thermal stability of catalytically active, transition metal oxide spinels is achieved by forming solid solutions with spinels with low catalytic activity and excellent surface area stability. Unsupported solid solutions of $CuCo_2O_4$, $CuMn_2O_4$, $CuFe_2O_4$, Co_3O_4, $ZnCo_2O_4$, or $CoFe_2O_4$ as the catalytically active spinel with $NiAl_2O_4$ or $ZnAl_2O_4$ as the high surface area spinel are characterized for physical properties, chemical structure, and catalytic activity. Oxidation, reduction, and water–gas shift reaction data were obtained for monolithically supported $CuMn_2O_4$ or $CuFe_2O_4$ in solid solution with $NiAl_2O_4$ or $ZnAl_2O_4$. $NiAl_2O_4$ appears to be exceptionally suitable as an isostructural host for catalytically active spinels. For various reasons, these solid solutions are doubtful candidates for use as automotive catalysts for oxidation or reduction.

Considerable attention has been given in recent years to catalytic treatment of automotive exhausts in order to meet federal emission standards for carbon monoxide, hydrocarbons, and oxides of nitrogen. One class of catalyst candidates which was extensively examined for oxidation of carbon monoxide and hydrocarbons is the transition metal oxides and their mixtures (1–14). Four modes of degradation of transition metal oxides used in treating automotive exhaust gases are thermal sintering, phase changes due to catalyst reactions with the support or support coating, oxidation–reduction degradation, and sulfur and other forms of poisoning.

[1] Present address: Shell Development Co., P.O. Box 100, Deer Park, Texas 77536.
[2] Present address: Engelhard Industries Division, Engelhard Minerals & Chemicals Corp., Menlo Park, Edison, N. J. 08811.

Transition metal ions dispersed in spinel matrices were studied for the decomposition of nitrous oxide (15, 16); however, there have been no reports on solid solutions of spinels for augmenting catalytic activity and stability in oxidation reactions. This is a report on work done with transition metal oxide spinels to minimize the first two forms of degradation through formation of solid solutions. Included are studies of various reaction processes involved in catalytic treatment of automotive exhaust gases.

Powdered samples of the solid solution catalysts were examined after varying thermal treatment for surface area stability and stability of catalytic activity in oxidation. Selected solid solutions were supported on monolithic substrates and examined for oxidation and reduction activity as a function of space velocity, water–gas shift activity, and susceptibility to poisoning by sulfur dioxide.

The first two forms of degradation can be suppressed by forming solid solutions of catalysts with certain isostructural compounds which possess high Brunauer–Emmett–Teller (BET) surface areas of superior thermal stability. However, these studies demonstrate that transition metal oxide solid solutions are doubtful candidates for use as automotive catalysts for a variety of reasons.

Experimental

Precursor solutions for individual spinel powder samples were prepared by dissolving in water and bringing to a total volume of 100 ml 0.1, 0.2, or 0.3 mole (with Co_3O_4) of the corresponding hydrated nitrates needed to produce the desired spinel stoichiometry. Concentrated ammonium hydroxide was added to these solutions, and the pH was adjusted to 7.0. The resultant hydroxide precipitate was washed thoroughly with distilled water and dried 16 hrs at 100°C; it was then fired at various temperatures for further evaluation. Solid solution precursors were prepared by adding equal amounts of two original precursor solutions and then precipitating, washing, drying, and firing in the same manner.

Celcor (Corning Glass Works) multicellular monolithic substrates, core-drilled to 2.54 cm, were used to support $CuFe_2O_4 \cdot NiAl_2O_4$, $CuFe_2O_4 \cdot ZnAl_2O_4$, $CuMn_2O_4 \cdot NiAl_2O_4$, and $CuMn_2O_4 \cdot ZnAl_2O_4$ solid solutions. The substrate is a porous cordierite ($2MgO-2Al_2O_3-5SiO_2$) ceramic which has a high use temperature, a high strength/weight ratio, and good thermal shock resistance. Porosity of the Celcor substrate is 35.8%; 35% of the pore volume is $> 10\mu$, and the mean pore diameter by volume is 6μ. Monolith channels have a square geometry with a cell count of 31 cells/cm^2, wall thickness of about 0.25 mm, and an open frontal area of approximately 70%.

Monoliths were impregnated with solid solution precursor, and excess solution was shaken from the monoliths. Mixed hydroxides were precipitated by passing ammonia vapor directly over the wet monoliths. Immediately after precipitation, the samples were fired 15 min at 600°C

Table I. Surface Area of Selected Spinels Heated in Air as a Function of Calcination Temperature

Spinel	Surface Area, m^2/g	
	Calcined 1 hr at 600°C	Calcined 1 hr at 600°C + 16 hrs at 800°C
Class I		
$CuCr_2O_4$	9.7	1.5
$CuMn_2O_4$	11	6
Co_3O_4	6.1	4.5
$CuFe_2O_4$	8	0.1
$ZnCo_2O_4$	16	3.3
$CoMn_2O_4$	12	5
$CoFe_2O_4$	17.9	0.1
$CuCo_2O_4$	17	0.3
Class II		
$NiAl_2O_4$	200	97 (43)[a]
$ZnAl_2O_4$	79	26 (16)[a]
$MgAl_2O_4$	125	91 (67)[a]

[a] Calcined 1 hr at 600°C + 16 hrs at 900°C.

to form oxides. This process was repeated 6–10 times until total loading was approximately 30 wt %. Catalyst-coated monoliths were then fired 16 hrs at 600°C.

Prior experience in our laboratories had indicated that the oxidation activity of catalyst candidates could be accurately appraised by using a differential scanning calorimeter (DSC) for powdered samples (*17*). Thus, all powdered samples were subjected to DSC analysis. A powdered sample in contact with a flowing reactant gas containing hexane or carbon monoxide and oxygen was heated, and the rise in sample temperature relative to that of an inert reference was measured. The temperature at which the oxidation exotherm was 50% of its maximum value was used

Table II. DSC Activity of Class I Spinels Heated in Air as a Function of Calcination Temperature

Spinel	DSC Activity, °C — Hexane/CO		
	Calcined 1 hr at 600°C	Calcined 1 hr at 600°C + 16 hrs at 800°C	Calcined 1 hr at 600°C + 24 hrs at 900°C
Class I			
$CuCr_2O_4$	271/199	310/238	332/266
$CuMn_2O_4$	216/171	277/221	321/271
Co_3O_4	204/135	257/171	288/238
$CuFe_2O_4$	266/199	371/316	>375/>425
$ZnCo_2O_4$	227/160	266/188	321/210
$CoMn_2O_4$	216/216	288/327	354/316
$CoFe_2O_4$	310/182	382/349	>375/>425
$CuCo_2O_4$	204/132	293/188	332/216

as the measure of activity. BET surface area and porosity and pore size distributions were determined and x-ray diffraction analyses were performed by the standard methods on the powdered samples. Catalyst distribution was determined by optical microscopy and surface roughness by scanning electron microscopy of the supported solid solutions.

Supported catalysts were evaluated for initial oxidation activity on a plug flow reactor at various space velocities using a standard reactant mix consisting of 1% carbon monoxide, 1.25% oxygen, 1000 ppm nitric oxide, and 250 ppm propylene (all on a dry basis), 10% water, and nitrogen the balance. This poison-free mix was used to simulate automotive exhaust conditions for lean operation. Temperature was varied from 100° to 600°C, and conversions of carbon monoxide and propylene were determined as a function of temperature. Space velocity was varied by using 1–4 monolithic pieces, each with a volume of 6 cm^3. At a total flow rate of 7500 cm^3/min, space velocities of 18,750, 25,000, 37,500, and 75,000/hr could be attained.

Table III. Surface Area of Copper Chromite–Nickel Aluminate Solid Solutions Heated in Air as a Function of Calcination Temperature

Spinel	Calcination Schedule, °C-hr	Surface Area, m^2/g
$CuCr_2O_4$	600-1	9.7
	600-1 + 800-16	1.5
	600-1 + 900-16	0.7
$3CuCr_2O_4 \cdot NiAl_2O_4$	600-1	38
	600-1 + 800-16	3.8
	600-1 + 900-16	2.1
$CuCr_2O_4 \cdot NiAl_2O_4$	600-1	74
	600-1 + 800-16	7.2
	600-1 + 900-16	5.9
$CuCr_2O_4 \cdot 3NiAl_2O_4$	600-1	130
	600-1 + 800-16	19
	600-1 + 900-16	6.9
$NiAl_2O_4$	600-1	200
	600-1 + 800-16	97
	600-1 + 900-16	43

Reduction activity was measured by using the same reactant mix except that oxygen was varied from 0.05 to 0.60% with the temperature held constant at either 482° or 593°C. Again space velocity was varied by using 1–3 monolithic units. Water–gas shift reaction was studied at 75,000/hr with the reactants being 1% CO, 10% H$_2$O, and N$_2$ the balance. For sulfur dioxide poisoning studies, either 50 or 100 ppm sulfur dioxide was added stepwise to the sample at 593°C with the oxygen level maintained at 0.40%; the effect on conversion was monitored continuously.

Carbon monoxide and nitric oxide were measured by nondispersive infrared analysis (NDIR), propylene by flame ionization detection (FID),

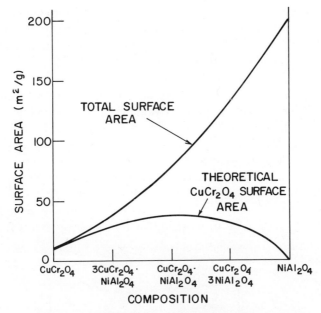

Figure 1. Total surface area and theoretical copper chromite surface area of samples fired 1 hr at 600°C as a function of composition

and oxygen by paramagnetic analysis. The ammonia which was produced was oxidized to nitric oxide over a monolithically supported platinum catalyst, and ammonia production was estimated from the differences between two nitric oxide NDIR detectors.

Results and Discussion

X-ray diffraction and microscopic studies revealed that calcined stoichiometric mixtures of coprecipitated hydroxides do indeed form spinels and solid solutions. With some mixtures, complete reaction was not always easily attained. For example, in the $CuO \cdot Fe_2O_3$ system, excess copper oxide and Fe_2O_3 peaks were found in x-ray diffraction patterns in addition to the major spinel phase. Calcined mixtures usually produced pure spinel compounds.

Eleven spinels were calcined in air at 800° and 900°C in order to reduce their surface areas. Two classes of spinels were defined (Tables I and II). Class I spinels generally had a high catalytic activity for oxidation of carbon monoxide and hydrocarbons, but they did not have or maintain a large surface area upon calcination. Class I spinels all had a surface area of less than 1 m^2/g after firing at 900°C for 16 hrs. Class II spinels, on the other hand, had high surface area and stability, but they were very poor oxidation catalysts.

The carbon monoxide and hexane oxidation activities of class I spinels were measured (Table II). Activity and activity loss were related to catalyst surface area and stability respectively. DSC data for class II spinels are not tabulated; these spinels were all essentially inactive with 50% exotherm temperatures being greater than 375°C.

Attempts were made to form solid solutions of class I and class II spinels in order to gain potential advantages of high activity, large surface area, and thermal stability through the composite. In initial studies, mole ratios of class I and class II spinels were varied. X-ray diffraction data verify solid solution for the complete range of compositions for these systems. BET surface area and DSC catalytic activity were measured for the various solid solutions as a function of calcination temperature. Data for the $CuCr_2O_4 \cdot NiAl_2O_4$ system are presented in Table III and Figures 1 and 2. The theoretical $CuCr_2O_4$ surface area was obtained

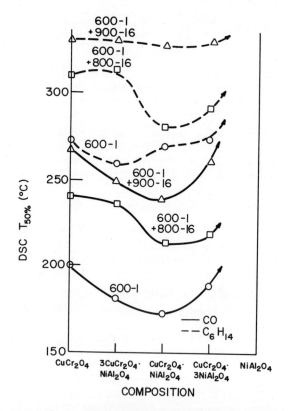

Figure 2. DSC activity as a function of calcination schedule and composition

Calcination schedule: °C-hr; all $NiAl_2O_4$ data were $>375°C$

Table IV. DSC Activity and Surface Area of Solid Solutions as a Function of Calcination Temperature

Class I Spinel	Calcination Schedule, °C-hrs	$NiAl_2O_4$ Solid Solution		$ZnAl_2O_4$ Solid Solution	
		Surface Area, m^2/g	DSC, °C C_6H_{14}/CO	Surface Area, m^2/g	DSC, °C C_6H_{14}/CO
$CuMn_2O_4$	600-1	124	210/127	90	182/121
	+ 800-16	8	266/171	2.6	177/221
Co_3O_4	600-1	74	243/154	67	254/160
	+ 800-16	20	232/138	11	249/149
$CuFe_2O_4$	600-1	54	299/149	61	257/154
	+ 800-16	16	310/171	5.2	335/243
$ZnCo_2O_4$	600-1	60	243/149	43.6	277/171
	+ 800-16	8	238/154	12.6	277/182
$CoMn_2O_4$	600-1	132	204/188	84.6	210/216
	+ 800-16	15	266/299	9.9	260/321
$CoFe_2O_4$	600-1	54	299/149	63.3	316/160
	+ 800-16	16	310/171	11.2	360/260
$CuCo_2O_4$	600-1	72	299/199	61.8	271/143
	+ 800-16	0.7	282/188	2.0	271/149

by multiplying the total (experimentally measured) surface area by the mole fraction of $CuCr_2O_4$. If the composite effects of surface area, surface area stability, and catalytic activity are realized, then the optimum system would be predicted at 50 mole % each class I and class II spinels (Figure 1). The data in Figure 2 verify this prediction.

A series of solid solutions (50 mole % each) was prepared from seven class I and two class II spinels. Each solid solution did stabilize catalytic activity and surface area (cf. Tables IV, I, and II). Furthermore, x-ray diffraction patterns revealed that class I spinels which did not easily form pure spinel phases readily formed a single spinel phase solid solution, e.g. the $CuO \cdot Fe_2O_3 \cdot NiO \cdot Al_2O_3$ system. Electron spectroscopy chemical analysis (ESCA) did not indicate any significant

Table V. Surface Area and Porosity of Celcor Substrates and Celcor Substrates Impregnated with Spinel Solid Solutions

System	Surface Area, m^2/g	Wall Porosity, %
Celcor substrate	<1	35.8
$CuFe_2O_4 \cdot NiAl_2O_4$ + substrate	6.5	38.8
$CuFe_2O_4 \cdot ZnAl_2O_4$ + substrate	10.9	39.6
$CuMn_2O_4 \cdot NiAl_2O_4$ + substrate	12.1	32.5
$CuMn_2O_4 \cdot ZnAl_2O_4$ + substrate	12.5	40.9

compositional differences between the ion-cleaned surface and the bulk (50 A of surface removed) for $CuFe_2O_4 \cdot NiAl_2O_4$.

Four of these solid solutions, supported on ceramic monolithic supports, were selected for further study with simulated automobile exhausts. These systems were chosen because they demonstrated good nitric oxide reduction activity and were among the best candidates for oxidation activity. The four systems consisted of $CuFe_2O_4$ and $CuMn_2O_4$ class I spinels in solid solution with $NiAl_2O_4$ and $ZnAl_2O_4$ class II spinels. Surface area and porosity of the monolithic support and of the support

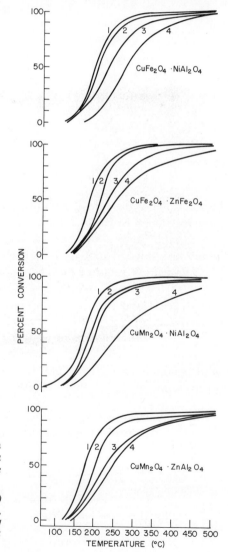

Figure 3. Carbon monoxide oxidation with various spinel solid solutions as a function of temperature and space velocity.

Reactants: 1% CO, 250 ppm C_3H_6, 1000 ppm NO, and 1.25% O_2 (all on dry basis), 10% H_2O, and balance N_2. Space velocity (hr^{-1}); 1, 18,750; 2, 25,000; 3, 37,500; and 4, 75,000.

impregnated with solid solutions were characterized (Table V). Differences in total porosity can be explained. The mercury porisimeter used is capable of measuring pore diameters as small as 200 A. Total porosity is a measure of the sum of monolith and catalyst porosity. When the catalyst pore size is less than 200 A and the catalyst is within the pores of the monolith, then the determined porosity could be less than that of the original monolith.

Broadening of the pore size distributions of monolith–catalyst systems relative to that of the bare monolith indicates that the pores of the

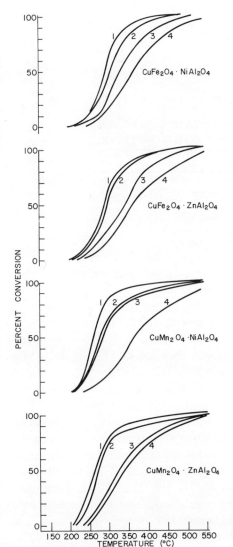

Figure 4. Propylene oxidation with various spinel solid solutions as a function of temperature and space velocity

Conditions same as in Figure 3

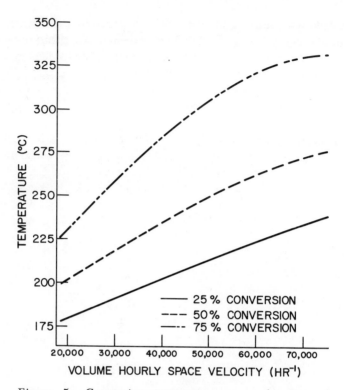

Figure 5. Conversion temperature as a function of space velocity for carbon monoxide oxidation with $CuFe_2O_4 \cdot NiAl_2O_4$

monolith are at least partially filled with catalyst. Microscopic examination of monolith–catalyst systems confirmed this.

The effect of increasing space velocity on the oxidation efficiency of these four catalysts is depicted in Figures 3 and 4. The findings were similar to those previously reported (18); conversion at a given temperature decreased rapidly as space velocity increased. This effect is also depicted in Figures 5 and 6. We cannot explain why the fall-off in conversion with $CuMn_2O_4 \cdot NiAl_2O_4$ was so much greater at higher space velocities.

Representative data from nitric oxide reduction studies are presented in Figures 7 and 8. Ammonia concentration decreased in all cases as the oxygen concentration was increased, and it was generally negligible at or beyond the stoichiometric value for oxygen (0.563%). In all cases ammonia production was greater at 482°C (900°F). Nitric oxide conversion decreased as the oxygen concentration approached stoichiometric for all 593°C (1100°F) runs; in some cases there was a slight increase in nitric oxide conversion as the oxygen level was increased above 0.05%.

For all 482°C runs, nitric oxide conversion decreased as the oxygen level was increased above 0.05%. Carbon monoxide and propylene conversions always increased with increasing oxygen concentrations. As would be expected, carbon monoxide, propylene, and nitric oxide conversions decreased as space velocity increased.

Data for propylene, carbon monoxide, and nitric oxide conversions and ammonia production with 0.25–0.55% oxygen are presented in Table VI. This range is termed the oxygen window, and it is representative of the oxygen range in an automobile equipped with a nitric oxide converter. Solid solutions with zinc aluminate produced much more ammonia than those with nickel aluminate whereas hydrocarbon conversion was definitely superior with nickel aluminate solid solutions. Carbon monoxide and nitric oxide conversions were approximately equivalent. As with the oxidation runs, $CuMn_2O_4 \cdot NiAl_2O_4$ was inferior to $CuFe_2O_4 \cdot NiAl_2O_4$ at high space velocities. Since any ammonia formed in a reducing convertor would be subsequently reconverted to nitric oxide in the oxidizing convertor, net nitric oxide conversion is very important. With at least 90% conversion required to meet federal emis-

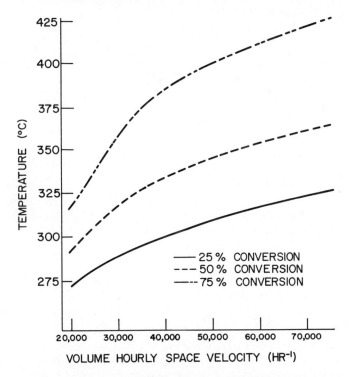

Figure 6. Conversion temperature as a function of space velocity for propylene oxidation with $CuFe_2O_4 \cdot NiAl_2O_4$

sion standards, none of these catalysts would be considered seriously as a reduction catalyst since 90% conversion was not achieved even at the relatively low space velocity of 75,000/hr and the reasonable temperature of 593°C.

It is of interest to determine whether NO is reduced to N_2 by reduction to NH_3 first, followed by decomposition into N_2 and H_2. NH_3 was indeed an intermediate with $CuMn_2O_4 \cdot NiAl_2O_4$, but the NH_3 concentration actually decreased with $CuMn_2O_4 \cdot ZnAl_2O_4$ as space velocity increased (Table VII). With the former catalyst, the nickel in the nickel aluminate apparently provided the ammonia decomposition function, whereas the latter catalyst was rather inactive for ammonia decomposition at 593°C. This finding agrees with those reported elsewhere (19). Findings were the same with $CuFe_2O_4 \cdot NiAl_2O_4$ and $CuFe_2O_4 \cdot ZnAl_2O_4$.

Since the water–gas shift reaction and perhaps the steam reforming reaction ($C_3H_6 + 3H_2O \rightleftarrows 3CO + 6H_2$) are the only sources of hydrogen in this reactant mix, it appears that at high space velocities hydrogen production *via* the water–gas shift reaction is retarded, and consequently

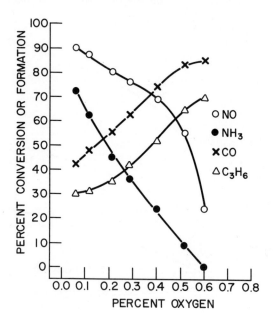

Figure 7. Conversion of NO, CO, and C_3H_6 and formation of NH_3 as a function of O_2 concentration with $CuFe_2O_4 \cdot NiAl_2O_4$ at 482°C and 75,000/hr space velocity

Reactants: 1% CO, 250 ppm C_3H_6, 1000 ppm NO (all on dry basis), variable O_2, 10% H_2O, and balance N_2

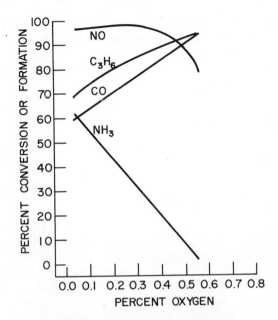

Figure 8. Conversion of NO, CO, and C_3H_6 and formation of NH_3 as a function of O_2 concentration with $CuFe_2O_4 \cdot NiAl_2O_4$ at 593°C and 75,000/hr space velocity

Average of four runs; reactants same as in Figure 7

less hydrogen but more carbon monoxide is available to react with nitric oxide. This effect is pronounced for $CuMn_2O_4 \cdot ZnAl_2O_4$ and $CuFe_2O_4 \cdot ZnAl_2O_4$. The data from water–gas shift studies with supported solid solutions are depicted in Figure 9. No real differences in ammonia production can be predicted on the basis of these data, with the possible exception that one would expect $CuFe_2O_4 \cdot ZnAl_2O_4$ to produce somewhat less NH_3 at 316°–482°C than $CuFe_2O_4 \cdot NiAl_2O_4$. Since the reverse was found, we assume that $CuFe_2O_4 \cdot NiAl_2O_4$ actually promoted reduction of NO to NH_3 much more rapidly.

When sulfur dioxide was introduced in poisoning studies, the change in conversion values was rapid in the first two minutes and slow thereafter. Poisoning was continued for 10 min; then SO_2 was removed from the reactants. Again change was rapid in the first two minutes, indicating some reversible adsorption. Thereafter there was slow recovery of activity. Similar behavior was reported with a copper chromite catalyst in an oxidizing atmosphere (20). Except for propylene oxidation over catalysts containing zinc aluminate, deactivation was always irreversible (Table VIII). Catalysts containing copper manganate were poisoned far more

Table VI. Propylene, Carbon Monoxide, and Nitric Oxide Conversions and Ammonia Formation within the Oxygen Window

Catalyst	VHSV, hr^{-1}	°C	Conversion, % HC	CO	NO	Net NO	NH_3, ppm
$CuFe_2O_4 \cdot NiAl_2O_4$	25,000	593	97	95	99	84	150
	37,500	593	96	92	96	74.5	215
	75,000	593	88	84	92	72.5	195
	75,000	482	43	71	67	43.5	235
$CuFe_2O_4 \cdot ZnAl_2O_4$	25,000	593	81	97	99	51.5	475
	37,500	593	79	89	95	61.5	335
	75,000	593	74	80	87	55	320
	75,000	482	42	72	69	34	350
$CuMn_2O_4 \cdot NiAl_2O_4$	25,000	593	94	94	96	86.5	95
	37,500	593	92	90	92	81	110
	75,000	593	80	81	80	69	110
	75,000	482	51	69	61	40.5	205
$CuMn_2O_4 \cdot ZnAl_2O_4$	25,000	593	75	89	95	60.5	345
	37,500	593	70	85	90	62.5	275
	75,000	593	69	88	87	62.5	245
	75,000	482	39	70	69	48	210

Table VII. Ammonia in Nitric Oxide Conversion with $CuMn_2O_4 \cdot X$ as a Function of Oxygen Concentration and Space Velocity

O_2, %	VHSV, hr^{-1}	NO, % converted $NiAl_2O_4/ZnAl_2O_4{}^a$	NH_3, ppm $NiAl_2O_4/ZnAl_2O_4{}^a$	NH_3, % NO converted $NiAl_2O_4/ZnAl_2O_4{}^a$
0.1	25,000	98.5/94	430/690	43.7/73.4
	37,500	95/96	440/690	46.3/71.9
	75,000	89/91	420/565	47.2/62.1
0.2	25,000	98.5/95	285/590	28.9/62.1
	37,500	95/96	325/580	34.2/60.4
	75,000	90.5/92	300/455	33.1/49.5
0.3	25,000	98/96	270/490	17.3/51.0
	37,500	95/96	200/450	21.1/46.9
	75,000	90/91	190/345	21.1/37.9
0.4	25,000	97/96	80/385	8.2/40.1
	37,500	93/93	95/300	10.2/32.2
	75,000	86/89	97/225	11.3/25.3

^{a} X in $CuMn_2O_4 \cdot X$.

readily than those containing copper ferrite, but much less ammonia was produced after poisoning. Water–gas shift catalytic sites were apparently severely poisoned. Carbon monoxide conversion with copper ferrite-containing catalysts was not affected as severely by sulfur dioxide, and consequently ammonia production was nearly the same after poisoning as before. Since all four catalysts would be rapidly poisoned by sulfur dioxide in an automobile exhaust environment, these catalysts cannot be considered for automotive applications unless convertor temperatures are maintained high enough to retard sulfate formation.

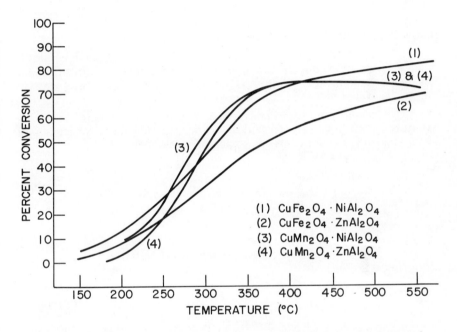

Figure 9. *Conversion of carbon monoxide as a function of temperature for the water–gas shift reaction at 75,000/hr space velocity*

Reaction: $CO + H_2O \rightleftarrows CO_2 + H_2$; reactants: 1% CO, 10% H_2O, and 89% N_2

Conclusions

From our interpretation of the data presented, we draw the following conclusions.

(a) Calcined stoichiometric mixtures of coprecipitated hydroxides form spinels and solid solutions.

(b) There are two classes of spinels: class I spinels are relatively active as oxidation catalysts, but they do not have or maintain a large surface area upon calcination; class II spinels do have high surface area and stability, but they are very poor oxidation catalysts.

(c) The surface composition of class I spinels appears to be directly proportional to their bulk composition when they are prepared in solid solution with class II spinels.

(d) Solid solution of class I spinels with class II spinels does in fact stabilize and promote catalytic activity.

(e) Conversions decrease rapidly at a given temperature as space velocity is increased.

(f) Solid solutions with nickel aluminate produce considerably less ammonia than solid solutions with zinc aluminate, apparently because nickel in the nickel aluminate provides an ammonia decomposition function.

(g) Hydrocarbon conversion with nickel aluminate is superior to that with zinc aluminate solid solutions under overall reducing conditions.

(h) Copper manganate-containing catalysts are poisoned by sulfur dioxide far more readily than copper ferrite-containing catalysts, but far less ammonia is produced after poisoning apparently because of poisoning of the water–gas shift catalytic sites.

(i) Carbon monoxide conversion with copper ferrite-containing catalysts is not severely affected by sulfur dioxide since poisoning of water–gas shift catalytic sites is apparently minimal.

Table VIII. Propylene, Carbon Monoxide, and Nitric Oxide Conversions (%) and Ammonia Formation as a Function of Sulfur Dioxide Poisoning[a]

Spinel	Gas	Before Exposure 50/100[b]	After 10-min Exposure 50/100[b]	After 8-min Recovery 50/100[b]
$CuFe_2O_4 \cdot NiAl_2O_4$	C_3H_6	92/90	82/80	82/80
	CO	88/86	78/76	84/83
	NO	96/99	78/80	79/90
	NH_3	225/260	250/250	300/320
$CuFe_2O_4 \cdot ZnAl_2O_4$	C_3H_6	61/65	70/76	69/67
	CO	69/74	58/58	61/64
	NO	85/91	70/65	81/84
	NH_3	430/500	370/415	415/470
$CuMn_2O_4 \cdot NiAl_2O_4$	C_3H_6	83/85	63/57	66/66
	CO	79/74	47/40	54/45
	NO	90/90	25/28	35/37
	NH_3	200/220	50/50	60/60
$CuMn_2O_4 \cdot ZnAl_2O_4$	C_3H_6	68/68	68/64	76/78
	CO	80/78	46/43	57/54
	NO	88/87	24/22	36/34
	NH_3	300/280	20/30	20/30

[a] O_2, 0.40% and $T = 593°C$; conversions given in % and ammonia formation in ppm.
[b] SO_2 in ppm.

(j) Since conversion with these catalysts is adversely affected by high space velocities, is severely poisoned by sulfur dioxide, results in excessive amounts of ammonia when nitric oxide is reduced, and is very sensitive to oxygen concentration under overall reducing conditions, the spinel solid solutions which we studied are doubtful candidates for use as automotive catalysts in either an oxidizing or reducing convertor.

Acknowledgments

The authors are indebted to Robert L. Johnson for his assistance in preparing samples and to John A. Myers for obtaining and analyzing many of the experimental data. The Physical Properties Research and Analytical Services Research Departments of Corning Glass Works generated many of the data essential to the successful completion of this study. The special attention of Brent Wedding and Norman A. Woodward in running the DSC and of Joel D. Sundquist in surface area analysis is greatly appreciated.

Literature Cited

1. Elovich, S. Y., Zhabrova, G. M., Margolis, L. Y., Roginskii, S. Z., *C.R. Acad. Sci. Russ.* (1946) **52**, 421.
2. Schachner, H., *Cobalt* (1959) **2**, 37.
3. Stein, K. C., Feenan, J. J., Thompson, G. P., Shults, J. F., Hofer, L. J. E., Anderson, R. B., *Ind. Eng. Chem.* (1960) **52**, 673.
4. Innes, W. B., Duffy, R., *J. Air Pollut. Control Ass.* (1961) **11**, 369.
5. Stein, K. C., Feenan, J. J., Hofer, L. J. E., Anderson, R. B., *U.S. Bur. Mines Bull.* (1962) **608**.
6. Dmuchovsky, B., Freerks, M. C., Zienty, F. B., *J. Catal.* (1965) **4**, 577.
7. Morooka, Y., Ozaki, A., *J. Catal.* (1966) **5**, 116.
8. Morooka, Y., Morikawa, Y., Ozaki, A., *J. Catal.* (1967) **7**, 23.
9. Andrushkevich, T. V., Boreskov, G. K., Popovskii, V. V., Muzykantov, V. S., Kimkhai, O. N., Sazanov, V. A., *Kinet. Katal.* (1968) **9** (3), 595.
10. Andrushkevich, T. V., Boreskov, G. K., Popovskii, V. V., Plyasova, L. M., Karakchiev, L. G., Ostan'kovich, A. A., *Kinet. Katal.* (1968) **9** (6), 1244.
11. Kimkhai, O. N., Popovskii, V. V., Boreskov, G. K., Andrushkevich, T. V., Dneprovskaya, T. B., *Kinet. Katal.* (1971) **12**, 371.
12. Artamonav, E. V., Sazonov, L. A., *Kinet. Katal.* (1971) **12**, 961.
13. Zhiznevskii, V. M., Fedevich, E. V., *Kinet. Katal.* (1971) **12**, 1209.
14. Boreskov, G. K., Proc. Congr. Catal., 5th (Miami Beach, Fla., August 1972) North-Holland, Amsterdam, paper **71**.
15. Cimino, A., Schiavello, M., *J. Catal.* (1971) **20**, 202.
16. Cimino, A., Pepe, F., *J. Catal.* (1972) **25**, 302.
17. Wedding, B., Farrauto, R. J., *Ind. Eng. Chem.* (1974) **13**, 45.
18. Schlatter, J. C., Klimisch, R. L., Taylor, K. C., *Science* (1973) **179**, 798.
19. Klimisch, R. L., Taylor, K. C., unpublished data.
20. Farrauto, R. J., Wedding, B., *J. Catal.* (1973) **33**, 249.

RECEIVED May 28, 1974.

14

Oxidation of CO and C_2H_4 by Base Metal Catalysts Prepared on Honeycomb Supports

J. T. KUMMER

Scientific Research Staff, Ford Motor Co., Dearborn, Mich. 48121

This report describes some laboratory experiments that were performed in order to assess the activity of base metal catalysts prepared using honeycombs as the substrate for the base metal catalyst alone or as the substrate for the washcoat together with the base metal catalyst. This investigation is continuing so this report constitutes only a progress report.

Since the substrate can carry only 10–30 wt % of the honeycomb as a washcoat (γ-Al_2O_3, ZrO_2, or other oxide), use of the honeycomb as a substrate places a greater demand on the activity of the base metal catalyst than the use of small (in order to avoid mass transfer problems) pellets which are entirely high area catalyst support. The honeycomb presents to the reacting gases only $\sim$1/3–1/5 the weight and corresponding surface area of the base metal catalysts as the same weight of small pellets. In actual practice, two honeycombs 4-11/16 inches in diameter and 3 in. long (2 lbs total) containing $\sim$0.4 lb of high surface area catalyst support are comparable with one 260-in.[3] pellet bed containing 3–6 lbs of high surface area catalyst support (pellets). Although the use of pellets in place of honeycombs can help to alleviate the problem of the lower specific catalytic activity of the base metals, it can also cause problems of attrition, warm-up, pressure drop, and space. Because of this, before pellets would be recommended for use with base metals, it is necessary to determine if it is possible to obtain sufficient base metal activity on a honeycomb of convenient size (<150 in.[3] for an $\sim$300-in.[3] engine).

We decided to evaluate the catalytic activity on a honeycomb of the best base metal catalyst that we know of in the absence of sulfur (a severe poison). If under the best conditions we could not provide sufficient catalytic activity (60% hydrocarbon oxidation activity in a typical vehicle installation after aging equivalent to 50,000 miles) with the honeycomb, then the use of pellets would seem necessary. If, on the other hand, we could provide sufficient activity under the best conditions, we would have only a partial answer since we would still have to devise some way to deal with the sulfur problem.

Table I. Relative Hydrocarbon Oxidation Activities of Various Oxides

Hydrocarbon	Relative Activity	Author	Ref.
Methane	$Co_3O_4 > NiO > MnO_2 > Fe_2O_3 > CuO$	Boreskov	(1)
Propane	$Co_3O_4 > Cr_2O_3 > MnO_2 > CuO > NiO$	Morooka et al.	(15)
Many	$Co_3O_4 > MnO_2 > NiO > Cr_2O_3 > FeO_3$	Stein et al.	(12)
Ethylene	$Co_3O_4 > Cr_2O_3 > MnO_2 > CuO$	Dmuchovsky et al.	(18)
Isobutylene	$Co_3O_4 > Mn_2O_3 > CuO > V_2O_5$	Zhiznevskii and Fedevich	(2)
Benzene	$Co_3O_4 > CuO > MnO_2 > Cr_2O_3$	Kimkhai et al.	(10)

Pure, unsupported transition metal oxides generally sinter badly at the high temperature (up to 1100°C) that occurs occasionally in auto exhaust. After heating to 1100°C, the residual surface area of the transition metal oxides is in the order of 1–2 m²/gram. Therefore, it appears necessary to use a high area washcoat (γ-Al_2O_3 or ZrO_2) or solid solutions more stable to sintering at high temperatures than the transition metal oxide itself in order to stabilize the surface area of the transition metal oxide at the high temperature. It is important that neither washcoat nor honeycomb react with and deactivate the active oxide.

There are considerable data in the literature on the activities of both single component base metal oxides and mixed oxides (1–22). Some of these data are summarized in Tables I, II, and III and in Figure 1.

Table II. Benzene Oxidation Activities of Various Oxides[a]

Metal	80% Conversion Temperature, °C
Co	163
Ni	270
Mn	282
Cu	317
Cr	337
Fe	345
Ti	365
Ce	370
Th	390
Al	417
W	480
Pb	496
V	565

[a] Data from Refs. 12 and 13.

Table III. Carbon Monoxide, Ethylene, and Ethane Oxidation Activities of Unsupported Catalysts[a]

Catalyst	Reaction Rate at 300°C,[b] cm³ CO_2/m² surface min		
	1% CO	0.1% C_2H_4	0.1% C_2H_6
Pd	500	100	1
Pt	100	12	1
Co_3O_4	80	0.6	0.05
$CuO \cdot Cr_2O_3$	40	0.8	0.02
Au	15	0.3	<0.2
MnO_2	3.4	0.04	
$LaCoO_3$	35	0.03	
CuO	45[c]	0.6	
Fe_2O_3	0.4	0.006	
Cr_2O_3	0.03	0.004	0.008
NiO	0.013	0.0007	0.0008

[a] Data from Ref. *23*.
[b] In presence of 1.0% O_2.
[c] At 250°C.

Under oxidizing conditions, transition metal oxides have valence states of 2–7. From published data on the oxidation of carbon monoxide and ethylene, the ratio R, where

$$R = \frac{\text{HC oxidation activity}}{\text{CO oxidation activity}}$$

increases as the valence of the transition metal oxide increases. Co^{2+} has poorer HC oxidation activity than CO oxidation activity whereas with V_2O_5 (V^{5+}) this ratio is higher. This observation concerns the ratios of activities and ignores the absolute oxidation activity per unit area which is extremely important. It is also apparent that the sulfates of the

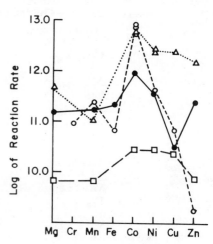

Figure 1. The rate of methane oxidation per unit surface area for various transition metal compounds (1)

○, pure oxides; △, colbaltites; □, ferrites; and ●, chromites

Proceedings of the Fifth International Congress on Catalysis higher valence oxides will decompose at lower temperatures than those of the lower valence oxides (Table IV and Figure 2) so that the sulfur tolerance of base metal catalysts will increase as the valence increases. In practice, vendors often add small amounts of Pd (100–500 ppm) to base metal catalysts in order to recover CO activity lost because of the presence of S.

Two materials with high catalytic activity are Co_3O_4 and $CuCr_2O_4$. Much of our work was with Co_3O_4, and it is based on preliminary work by Bettman (25) and Yao (21). They demonstrated that impregnating γ-Al_2O_3 with $Co(NO_3)_2$ yields a very poor catalyst which has a much lower activity for oxidation of CO and ethylene than does pure Co_3O_4 (per unit surface area of Co_3O_4). Use of ZrO_2 as a catalyst support, however, yields a catalyst in which the oxidation activity per unit area of Co_3O_4 is largely maintained.

Honeycomb Parameters Required

This report is concerned with the preparation and testing of small honeycomb catalysts with Co_3O_4 or copper chromite as the active phase. With Co_3O_4, the oxidation of C_2H_4 is one of the more difficult reactions to catalyze. Both CO and the higher molecular weight hydrocarbons (unburned fuel) are more readily oxidized (21). C_2H_4 is one of the main constituents of cracked fuel from the cylinder quench layer.

Yao's findings for C_2H_4 oxidation with Co_3O_4 (21) can be expressed by the equation:

$$\frac{\text{cm}^3\ C_2H_4\ \text{oxidized}}{\text{min m}^2} = 0.85 \times 10^7\ e^{-\frac{20,000}{RT}}\ P_{C_2H_4}^{1/2}\ P_{O_2}^{1/2}\ P_{H_2O}^{-1/5} \quad (1)$$

where the pressures are expressed in mm Hg. This equation can be used to calculate the surface area of Co_3O_4 (per cm³ honeycomb) required to produce a given percent conversion of C_2H_4 in the inlet gas to the honeycomb at a given space velocity and temperature. For the conditions: ethylene concentration, 0.1% in the inlet gas to the honeycomb (330

Table IV. Ease of Decomposition of Various Transition Metal Sulfates[a]

Salt	Temperature, °C
$MnSO_4$	649
$NiSO_4$	600
$CoSO_4$	590
$CuSO_4$	500
$Cr_2(SO_4)_3$	370
$Fe_2(SO_4)_3$	315

[a] Decomposition in vacuum; data from Ref. 24.

ppm hexane); O_2, 2%; water, 10%; space velocity, 80,000/hr; and no CO to produce a temperature increase in the honeycomb (isothermal conditions), we calculated the temperature of 50% conversion as a function of Co_3O_4 surface area (m^2/cm^3 honeycomb) (Figure 3). The temperature of 50% conversion for CO would have occurred at much lower

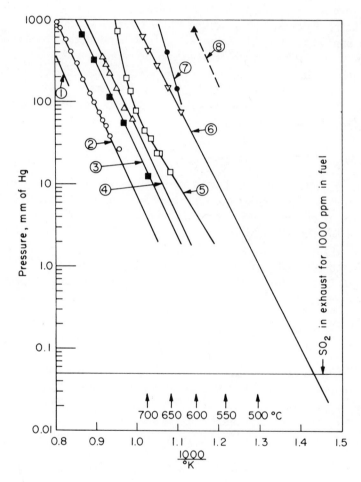

Figure 2. Total pressure of decomposition (to $SO_3 + SO_2 + O_2$) of base metal sulfates

1. $MnSO_4 \rightarrow Mn_3O_4 + xSO_3 + y(SO_2 + \frac{1}{2}O_2)$
2. $CoSO_4 \rightarrow CoO + xSO_3 + y(SO_2 + \frac{1}{2}O_2)$
3. $NiSO_4 \rightarrow NiO + xSO_3 + y(SO_2 + \frac{1}{2}O_2)$
4. $2CuO \cdot SO_3 \rightarrow 2CuO + xSO_3 + y(SO_2 + \frac{1}{2}O_2)$
5. $Al_2(SO_4)_3 \rightarrow Al_2O_3 + xSO_3 + y(SO_2 + \frac{1}{2}O_2)$
6. $Fe_2(SO_4)_3 \rightarrow Fe_2O_3 + xSO_3 + y(SO_2 + \frac{1}{2}O_2)$
7. $2Cr_2O_3 \cdot 3SO_3 \rightarrow 2Cr_2O_3 + xSO_3 + y(SO_2 + \frac{1}{2}O_2)$
8. $2VOSO_4 \rightarrow V_2O_5 + xSO_3 + y(SO_2 + \frac{1}{2}O_2)$

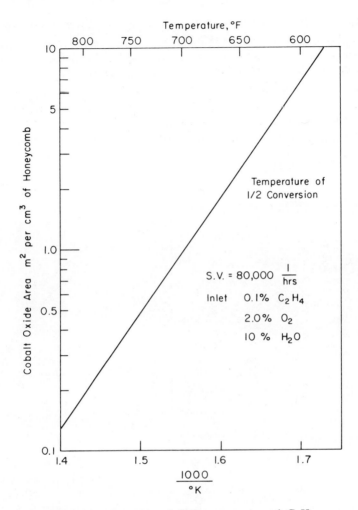

Figure 3. Temperature of 50% conversion of C_2H_4 as a function of cobalt oxide surface area on honeycomb

temperatures and would have increased the catalyst temperature above inlet temperature ∼89°C for every 1% CO in the inlet if we had assumed that CO was present.

For good cold-start hydrocarbon removal (*see* Figure 3), Co_3O_4 surface areas should be in the order of 10 m^2/cm^3 honeycomb (or ∼17 m^2/g honeycomb if honeycomb density is 0.6 g/cm^3). These values are difficult to obtain. The surface area of pure Co_3O_4 heated to the low temperature of 315°C is equivalent to ∼25 m^2/g. If all the voids of the honeycomb (0.15 cm^3/g) were filled with Co_3O_4 with a density of 2 g/cm^3, the Co_3O_4 surface area would be ∼7.5 m^2/g honeycomb (23 wt

% Co_3O_4). On heating to 982°C, however, the Co_3O_4 surface area would decrease to 1–2 m^2/g Co_3O_4 (0.3–0.6 m^2/g honeycomb). The temperature of 50% conversion for ethylene (Figure 3) is ~400°C which is high. Therefore, it is desirable to use a washcoat with a high surface area which is thermally stable toward sintering and chemically unreactive toward Co_3O_4 in order to preserve Co_3O_4 surface area at high temperatures. In addition, presence of the washcoat will help to isolate the Co_3O_4 from the honeycomb which may contain constituents that can poison the catalyst.

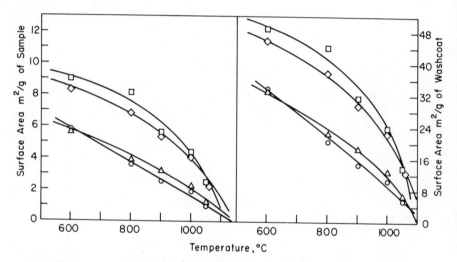

Figure 4. Sintering of ZrO_2 washcoat. Data from Goodsel (26).
Sintering: 5 hrs in air + 10% water vapor; total weight of washcoat on honeycomb, ~18 wt %; □, ZrO_2 + 20 mole % SiO_2; ◇, ZrO_2 + 11.4 mole % SiO_2; △, ZrO_2 + 5 mole % SiO_2; and ○, pure ZrO_2

Unfortunately, the commonly used washcoat Al_2O_3 is reactive toward Co_3O_4 although it can be used in ways that will minimize this. ZrO_2 appears to be unreactive when pure. Its surface area as a function of temperature in 10% H_2O vapor was determined by Goodsel (26) (see Figure 4); he also demonstrated that adding a small amount of SiO_2 stabilizes the surface area of ZrO_2. After 5-hrs heating at 1010°C, the surface area of a honeycomb containing 18% ZrO_2 + 5% SiO_2 was ~4 m^2/g honeycomb (2.5 m^2/cm^3). If this area could be covered with Co_3O_4 (2.5 m^2 Co_3O_4 surface area/cm^3 honeycomb) which is the maximum dispersion of Co_3O_4 we could expect and which we have not attained, then the temperature of 50% conversion of ethylene would be 343°C (Figure 3).

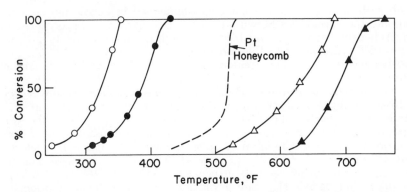

Figure 5. CO and C_2H_4 oxidation activity of American Lava honeycomb 795 coated with 6% Co_3O_4

Honeycomb pretreated at 351°C; space velocity, 80,000/hr; ○, 1% CO + 2% O_2; △, 0.2% C_2H_4 + 2% O_2; open symbols, dry inlet; solid symbols, wet inlet (3% H_2O); and – – – –, coincident curves for CO and C_2H_4 oxidation

Experimental

The principal substrate was the American Lava honeycomb, and the principal washcoat was colloidal zirconia (National Lead Industries) (~20 wt % solids). The honeycomb was immersed in colloidal zirconia, the excess was blown out the passageways with compressed air, and the honeycomb was then dried with a hot air gun. This process was repeated four times to produce a ZrO_2 washcoat of 20–25% wt %. The washcoat was usually heated to 815°C for 24 hrs before further use. The honey-

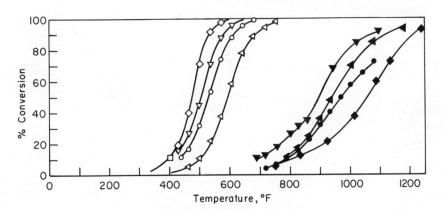

Figure 6. Oxidation of CO and C_2H_4 on American Lava 795 honeycomb containing 20% ZrO_2 + 6% Co_3O_4

▽, heated at 800°C in air 24 hrs; ○, Li_2O (4 mole % of Co_3O_4) added, heated at 800°C; ◇, Na_2O (4 mole % of Co_3O_4) added, heated at 800°C; ◁, Cu_2O (4 mole % of Co_3O_4) added, heated at 800°C; open symbols: inlet: 1% CO, 2% O_2, and 3% H_2O; and solid symbols: inlet: 0.2% C_2H_4, 2% O_2, and 3% H_2O

comb with or without a washcoat was immersed in cobalt nitrate solution, and the excess was blown out of the passageways with compressed air. Anhydrous ammonia was blown through the honeycomb to precipitate cobalt hydroxide, and the honeycomb was dried with a hot air gun. The honeycomb was then treated at 315°, 649°, or 815°C for 24 hrs before use.

For activity measurements, the honeycomb was crushed and sieved to 16–32 mesh. About 100 mg of this material was tested for catalytic activity in a quartz tube (~2.5 mm inner diameter). The tube was filled to ~4 cm; the flow rate was ~230 cm^3/min (standard temperature and pressure). For CO activity measurement, the nominal inlet gas composition was 1% CO, 2% O_2, 3% H_2O (or dry), and 94% He; for ethylene activity measurements, it was 0.2% ethylene, 2% O_2, 3% H_2O (or dry), and 94.8% He. Sulfur dioxide was added from a premixed tank of Ar–SO_2 (1000 ppm SO_2). The exit gas from the catalytic reactor was analyzed by an on-stream mass spectrometer. Space velocity based on quartz tube volume was ~70,000/hr whereas that based on the honeycomb volume equivalent to the sample weight was ~80,000/hr. During the constant volume sampling (CVS) test, the space velocities encountered in the vehicle (1970 Galaxie, 351-in.^{3} engine, 17% exhaust gas recirculation) were 30,000–120,000/hr with the larger values occurring only for short times during the large acceleration; the average value was ~60,000/hr. These values are for a PTX5 honeycomb (Englehard Industries) 3 in. long, one for each four cylinders.

We observed that by using the same ZrO_2 solution and the same copper–chromium solution we could prepare successive honeycomb catalysts with which the temperature for 50% conversion of CO and HC varied less than ±10°C.

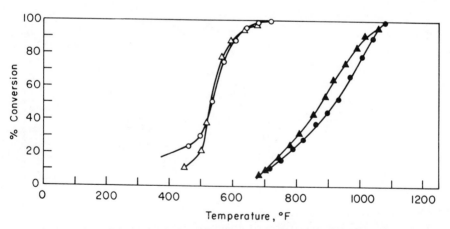

Figure 7. Oxidation activity of American Lava honeycomb 795 coated with ZrO_2 (20 wt %) and either Co_3O_4 or copper chromite

Honeycomb preheated 24 hrs at 815°C before use; space velocity, 80,000/hr; ○, 6 wt % Co_3O_4; △, 13 wt % copper chromite; open symbols: inlet: 1% CO, 2% O_2, 3% H_2O; and solid symbols: inlet: 0.2% C_2H_4, 2% O_2, and 3% H_2O

Table V. Activity Tests on Honeycombs[a]

Sample	CO Oxidation, $T_{50}°C$[b]		Butene Oxidation, $T_{50}°C$[b]	
	Fresh	Aged[c]	Fresh	Aged[c]
Honeycomb + 21% ZrO_2 + 3.1% Co_3O_4	155	165	240	255
Honeycomb + 23% ZrO_2 + 3.0% Co_3O_4 + 300 ppm Pd	165	170	230	260
Honeycomb + 15% (ZrO_2 + 1/6 Cr_2O_3) + 3.1% Co_3O_4	145	180	245	245

[a] Honeycombs: American Lava 795; space velocity: 10,000/hr; reactants: 1% CO + 1% O_2 or 1000 ppm butene + 1% O_2.
[b] T_{50} = temperature of 50% conversion.
[c] Aged samples were treated 1 wk at 800°C in 10% water vapor.

Results

Numerous screening tests were run. Typical data are presented in Figures 5, 6, and 7 and in Table V. Some of the findings can be summarized as follows.

First, the activity of base metal honeycombs tested for CO oxidation was good with the temperature of 50% conversion at 80,000/hr space velocity being about 260°C after aging 16 hrs at 815°C. This is comparable to the activity of honeycombs containing Pt that are presently being used.

Second, the hydrocarbon activity as judged by C_2H_4 oxidation was inferior to that of Pt honeycombs. With the base metal honeycomb, the temperature of 50% conversion at 80,000/hr space velocity was ∼482°C whereas with the Pt honeycombs it was ∼260°C, the same as for CO oxidation. The temperature for 50% conversion of other hydrocarbons (aromatic and large molecular weight olefins and aliphatics) would be lower.

Third, the findings for copper chromite supported on ZrO_2 were similar to those for Co_3O_4 on ZrO_2 (Table III). Addition of Cr_2O_3 to the Co_3O_4 increased the HC activity and decreased the CO activity after aging (Table V).

Fourth, use of ZrO_2 as the washcoat appeared to allow the use of Co_3O_4 and copper chromite without deactivation. However, the total surface area of the Co_3O_4 with this washcoat after high temperature treatment was not as large as desirable in order to give good low temperature HC oxidation activity. In addition, we found (data not presented) that incorporation of 5% SiO_2 into the washcoat increased surface area retention but deactivated Co_3O_4.

It was reported in the literature and also demonstrated in this laboratory that both Co_3O_4 and copper chromite are poisoned by sulfur. This results from the accumulation of sulfate groups on the catalyst surface. The base metal sulfates and aluminum sulfate are very stable, and they decomposed to the oxide only at temperatures above $\sim$650°C (see Table IV and Figure 2). Above 650°C, activity was restored because of sulfate decomposition. When a base metal catalyst was subjected to high temperatures before being cooled down for a CVS test, it had good activity for a short period of time which was dependent on the sulfur content of the gasoline and the surface areas of the washcoat and base metal catalyst.

With 30 standard ft^3/min (scfm) as the average exhaust flow per catalyst unit (one honeycomb) and a gasoline sulfur content of 300 ppm (1 g S/gal), the exhaust flow presents to the catalyst $\sim$44 mg SO_2/min. A 500-g honeycomb with 6% Co_3O_4 would contain 30 g Co_3O_4, and, if its surface area were 5 m^2/g Co_3O_4 ($\sim$0.18 m^2/cm^3 honeycomb), a monolayer of SO_4 groups would amount to $\sim$80 mg. During the CVS test, the exhaust gas would carry enough SO_2 to cover one-half the cobalt oxide surface in the first minute and to cover it completely in 2 min. It is expected, however, that the first part of the catalyst will consume more than a monolayer of SO_2 and consequently delay the poisoning of the remainder of the bed. It is not known how much SO_2 the ZrO_2 washcoat can absorb, but, whatever the amount, it will delay poisoning of the Co_3O_4 surface. In this example, the time for deactivation can be in-

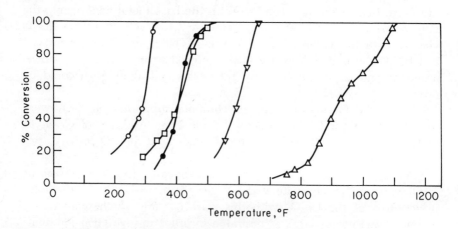

Figure 8. CO oxidation by Co_3O_4 (4%) on ZrO_2

Bulk density, 1.3 g/cm^3; space velocity, 80,000/hr; inlet: 1% CO and 2% O_2; ○, dry inlet; ●, 3% H_2O; △, 3% H_2O + 66 ppm SO_2; ▽, 3% H_2O, regenerated at 649°C 16 hrs; and □, 3% H_2O, regenerated at 799°C 2 hrs

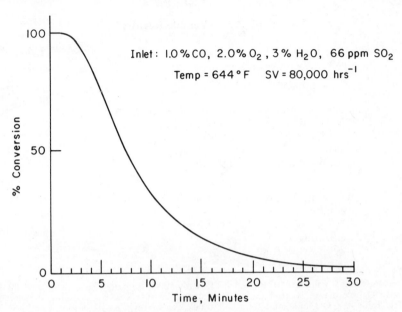

Figure 9. Decrease in the oxidation of CO upon addition of 66 ppm of SO_2 to the inlet stream as a function of time

creased by a larger base metal oxide surface, but, as noted above, with honeycomb substrates we are limited in this respect.

The CVS test findings depend on whether the honeycomb has been subjected to idle conditions (the 427°–538°C temperature is not high enough to remove the S from the oxide surface) or to high temperatures (>700°C) which remove the S before the test is run. In the test, the catalyst should reach a temperature >700°C quickly (in several minutes) in order to maintain its activity if it started out free from sulfur.

There could be some compatibility problems with the NO_x catalyst, if the NO_x catalyst did not retain S but did retard the heating of the oxidation catalyst. Large pellet beds (260 in.³) could be advantageous because a much larger base metal surface area would be available since most of the bed is high area support. As stated previously, the cobalt surface area for a 3–6 lb bed would be ~10 times that for the honeycomb, and the time for heat-up would be only about twice as long.

The behavior of a Co_3O_4 catalyst supported on ZrO_2 is depicted in Figure 8. The ZrO_2–Co_3O_4 (4% Co_3O_4) was used as such without deposition on the honeycomb support. The total surface area was 29 m²/g; that of the Co_3O_4 was estimated at 1.5–7 m²/g. The plot of CO oxidation in the presence of 66 ppm SO_2 is the equilibrium curve because sufficient time was allowed at each point for equilibrium to be reached. Regenera-

Table VI. Vehicle Results[a]

Test	Catalyst	Conditions	Conversion, %			
			CVS/CH		CVS/H	
			CO	HC	CO	HC
I	one 50-in.[3] honeycomb (6% Co_3O_4, 20% ZrO_2) for total exhaust	after 10 miles on fuel containing 200 ppm S	2	0		
		after subsequent 150 miles on S-free fuel	25	24		
II	two 50-in.[3] honeycombs in series (15% copper chromite, 20% ZrO_2) for total exhaust	S-free fuel	67	60	80	64
		fuel containing 200 ppm S	42	39	43	42

[a] Conditions: vehicle: 1971 Ford LTD with 429-in.[3] engine; catalyst: aged 30 hrs at 815°C before use, located under driver; and $T = \sim 482$°C. Baseline values: 2.7 g HC/mile, 38 g CO/mile, and 1.4 g NO/mile. Testing: CVS with cold (C) and hot (H) starts according to federal test procedures.

tion was effected by a flow of 1% O_2–0.2% C_2H_4 in the absence of SO_2 at a 80,000/hr space velocity at the indicated temperature. Regeneration was partial at 649°C and complete at 815°C. Figure 9 depicts the activity of a fresh portion of the catalyst used above when it was exposed at 340°C to 1% CO, 2.0% O_2, 3% H_2O, and 66 ppm SO_2 (1000 ppm in fuel) at a space velocity of 80,000/hr. If the cobalt oxide area is 7 m^2/g catalyst, then in 10 min the inlet gas supplies 2.8 mg SO_2/g catalyst or enough

Table VII. Vehicle Results[a]

Test	S in fuel, ppm	HC		CO	
		g/mile	% convtd.	g/mile	% convtd.
		CVS/H			
I	144	0.61	70	5.7	82
II	144	0.68	67	6.3	80
III	144	0.63	69	6.0	81
		CVS/CH			
III	144	0.82	67	7.0	82

[a] Vehicle: 1971 Ford LTD with 429-in.[3] engine; catalyst located ~4 in. from manifold flange; one honeycomb per bank of four cylinders (total, 2); honeycomb: 4-11/16 in. in diameter and 4.5 in. long with 66 in.[3] exposed to exhaust gas; and catalyst inlet temperatures: ~371°C in 40 sec, ~499°C in 80 sec, and ~732°C maximum during big acceleration, with mean of ~566°–593°C. Baseline values for CVS/CH: 2.5 g HC/mile, 40.0 g CO/mile, and 1.25 g NO/mile; and for CVS/H: 2.06 g HC/mile and 31.0 g CO/mile. Catalyst: copper chromite–ZrO_2 on American Lava honeycomb, heated 16 hrs at 815°C before use.

to cover ~80% of the cobalt oxide area. The shape of the curve in Figure 9 would imply that all the entering SO_2 remains on the cobalt surface area.

Vehicle Results

Vehicle results are presented in Table VI. The use of one 50-in.³ honeycomb for the total exhaust is standard procedure for evaluating vendors' noble metal catalysts. The base metal catalysts were aged 30 hrs at 815°C before use, and the activity represents low mileage activity. The catalyst temperature in this vehicle during the CVS cycle (427°–538°C) was low for base metal applications, particularly with S-containing fuels. Vehicle results with two larger honeycombs, each used for four cylinders, mounted closer to the engine are presented in Table VII. Honeycombs, which can more readily be used closer to the exhaust manifold and consequently hotter, are more advantageous than pellet beds in alleviating the effect of SO_2 deactivation.

Conclusion

The 3-in. long honeycomb with cobalt oxide as the active catalyst appears to be too sensitive to sulfur poisoning to be useful.

The activity of the 4.5-in. honeycombs with copper chromite as the active catalyst, when mounted 4 in. from the exhaust flanges of a vehicle and when used with low sulfur fuel (144 ppm S), appears to be somewhat less than that of a Pt honeycomb (for which we would expect >90% CO conversion and >80% HC conversion in the tests of Table VII).

Work is continuing in an effort to find catalysts with sufficient activity and resistance to sulfur poisoning. No durability data in the presence of sulfur have been obtained as yet.

Literature Cited

1. Boreskov, G. K., "Catalytic Activity of Transition Metal Compounds in Oxidation Reactions," Proc. Intern. Congr. Catal., 5th (Miami, 1972) North-Holland, Amsterdam, paper **71**.
2. Zhiznevskii, V. M., Fedevich, E. V., "Investigation of the Activity of Metal Oxides in the Oxidation of Isobutylene," *Kinet. Katal.* (1971) **12**, 1209.
3. Artamonav, E. V., Sazonov, L. A., "Catalytic Activity of Oxides of Rare Earth Elements in the Oxidation of Carbon Monoxide," *Kinet. Katal.* (1971) **12**, 961.
4. Klier, K., "Oxidation and Reduction Potentials and Their Relation to the Catalytic Activity of Transition Metal Oxides," *J. Catal.* (1967) **8**, 14.
5. Roginskii, S. Z., Al'tshuler, O. V., Vinogradova, O. M., Seleznev, V. A., Tsitovskaya, I. L., "Catalytic Properties of Zeolites Containing Transition Metal Ions in Oxidation Reduction Reactions," *Dokl. Akad. Nauk SSSR* (1971) **196**, 872.

6. Ganieva, T. F., Belenkii, M. S., Sultanov, M. Yu., "Investigation of the Copper Chromite Alumina Catalyst," *Kinet. Katal.* (1970) **11**, 1196.
7. Dwyer, F. G., "Catalysis for Control of Automotive Emissions," *Catal. Rev.* (1972) **6**, 261.
8. Yolles, R. S., Wise, H., "Catalytic Control of Automotive Exhaust Emissions," *Critical Rev. Environ. Control* (1971) **2**, 125.
9. Rudham, R., Sanders, M. K., "The Catalytic Properties of Zeolite X Containing Transition Metal Ions for Methane Oxidation," *J. Catal.* (1972) **27**, 287.
10. Kimkhai, O. N., Popovskii, U. V., Boreskov, G. K., Andrushkevich, T. V., Dneprovskaya, T. B., "Catalytic Properties of Group IV Metal Oxides in Oxidation Reactions III Benzene," *Kinet. Katal.* (1971) **12**, 371.
11. Schachner, H., "Cobalt Oxides as Catalysts for the Complete Combustion of Automotive Exhaust Gases," *Cobalt* (1959) **2**, 37.
12. Stein, K. C., Feenan, J. J., Thompson, G. P., Shults, J. F., Hofer, L. J. E., Anderson, R. B., "Catalytic Oxidation of Hydrocarbons," *Ind. Eng. Chem.* (1960) **52**, 673.
13. Stein, K. C., Feenan, J. J., Hofer, L. J. E., Anderson, R. B., "Catalytic Oxidation of Hydrocarbons," U.S. Bur. Mines Bull. (1962) **608**.
14. Morooka, Y., Ozaki, A., "Regularities in Catalytic Properties of Metal Oxides in Propylene Oxidation," *J. Catal.* (1966) **5**, 116.
15. Morooka, Y., Morikawa, Y., Ozaki, A., "Regularity in the Catalytic Properties of Metal Oxides in Hydrocarbon Oxidation," *J. Catal.* (1967) **7**, 23.
16. Mochida, I., Hayata, S., Kato, A., Seiyama, T., "Catalytic Oxidation over Molecular Sieves Ion Exchanged with Transition Metal Ions," *J. Catal.* (1971) **23**, 31.
17. Takahashi, T., Jotani, F., "Metal Catalysts for Catalytic Combustion," *Fuel Soc. Jap. J.* (1967) **46**, 828.
18. Dmuchovsky, B., Freerks, M. C., Zienty, F. B., "Metal Oxide Activities in the Oxidation of Ethylene," *J. Catal.* (1965) **4**, 577.
19. Innes, W. B., Duffy, R., "Exhaust Gas Oxidation on Vanadia Alumina Catalyst," *J. Air Pollut. Control Ass.* (1961) **11**, 369.
20. Klimisch, R. M., "Oxidation of CO and Hydrocarbons over Supported Transition Metal Oxide Catalysts," Proc. Nat. Symp. Heterogen. Catal. Control Air Pollut., 1st (Philadelphia, November, 1963).
21. Yao, Y., "The Oxidation of Hydrocarbons and CO over Metal Oxides. III. Co_3O_4," *J. Catal.* (1974) **33**, 108.
22. Hofer, J. C., et al., "Specificity of Catalysts for the Oxidation of $CO-C_2H_4$ Mixtures," *J. Catal.* (1964) **3**, 451.
23. Yao, Y., unpublished data.
24. Cueilleron, J., Hartmanshenn, O., *Bull. Soc. Chim. France* (1959) 168.
25. Bettman, M., unpublished data.
26. Goodsel, A., unpublished data.

RECEIVED August 12, 1974.

INDEX

A

Acidity, effects of the support on	35
Activated oxides with carriers, reactions of	159
Activity	
catalyst	133
hydrocarbon	187
oxidation	163
profile	111
reduction	164
Activities, carbon monoxide, ethylene, and ethane oxidation	180
Activities of various oxides, hydrocarbon oxidation	179
Agency (EPA) requirements, Environmental Protection	144
Aging	
catalysts, synthetic means of	145
of catalysts, engine	3
on NO reduction, effect of catalyst	159
palladium and platinum resistance to sulfur and thermal	139
Air Act, Clean	1
Air using vanadia–alumina catalyst, interaction of NO with NH_3 in	15
Alumina(s)	
carriers	148
catalyst, interaction of NO with NH_3 in air using vanadia	15
chemically pure	150
lattice structure of transition	153
metastability of transition	157
remedies for instability of	155
stability of various	153
Aluminum hydroxide to corundum, transformation of an	152
Aluminum molybdate	157
Aluminate, copper chromite–nickel	164
Ammonia	17, 48, 170
analysis capability for	23
–nitric oxide reaction stoichiometry	37
Ammonia selectivity	32, 36
Analysis capability for ammonia	23
Analysis, x-ray diffraction analysis	88
Areas, surface	183
Arrhenius plot for a poisoned catalyst	111
Automotive	
catalysis, distribution of catalytic material on support layers in	116
Automotive *(continued)*	
emissions catalyst test unit, schematic	134
emission control catalysts, degradation in	103
emission control catalysts, oxidative	133
system, definition of	24
Automobile exhaust	
deoxiding temperatures for metals in	51
desulfiding temperature in	48
equilibrium composition of	45
gas, catalytic reduction of NO_x emissions in	244
NO_x catalysts	85
thermodynamic interaction between transition metals and	39
transition metals in	49

B

Base metal	
catalyst, effect of catalyst thickness on CO conversion	126
catalysts, oxidation of CO and $C_2H_4^-$ by	178
–noble metal catalyst	96
oxidation catalysts, comparison of platinum and	72
pellets for use with	178
vs. platinum catalysts	83
Bed dilution, effect of catalyst	141
1-Butene oxidation	19

C

Calcination	151, 165
Calcium	85, 101
Carbon monoxide	164, 171
emissions	30
ethylene, and ethane oxidation	180
oxidation	137
Carburetor system, sensor–	3
Carriers	
alumina	148
properties of	149
reactions of activated oxides with	159
thermally stable catalyst	147
Catalysis, distribution of catalytic material on support layers in automotive	116
Catalysis, NO_x	51

Catalytic
converters 74
material on support layers in automotive catalysis, distribution of 116
reduction of NO_x emissions in auto exhaust gas 244
Catalyst
activity 133
aging on NO reduction, effect of 159
Arrhenius plot for a poisoned .. 111
base metal–noble metal 96
bed dilution, effect of 141
carriers, thermally stable 147
copper–nickel extrudate 88
composition on catalyst durability, effect of 27
degradation by poisons 85
deterioration, change in catalyst volume on 108
durability, effect of catalyst composition on 27
effectiveness of a 39
effect of catalyst thickness on CO conversion by base metal .. 126
for exhaust emission control, platinum 54
inlet temperature 8
interaction on NO with NH_3 in air using vanadia–alumina 15
layer thickness on CO conversion for platinum kinetics, effect of 127
NO_x 39, 189
reduction 1
poisoning 107
by lead and phosphorus compounds 54
properties 74
poison deposition on the 68
poisons, oil, and fuel additives as 57
porosity, monolith 169
selectivity 32
sphere, metals distribution within the 141
supports, properties of 33
supported ruthenium 2
temperature on lead poisoning, effect of 65
test unit, schematic of automotive emissions 134
thickness on CO conversion by base metal catalyst, effect of 126
three-way conversion (TWC) .. 3
warm-up characteristic of 8
volume on catalyst deterioration, change in 108
Catalysts
automobile exhaust NO_x 85
base metal vs. platinum 83
by contamination, degradation of NO_x 85
comparison of platinum and base metal oxidation 72

Catalysts (continued)
degradation in automotive emission control 103
distribution of contaminants on monolithic NO_x 95
engine aging of 3
monolithic 32
NO_x removal 32
oxidative automotive emission control 133
oxidation of CO and C_2H_4 by base metal 178
Catalysts, palladium 106, 114
platinum 144
monolithic oxidation 54
poisoning of platinum 62
preparation of monolithic catalysts 25
preparation of spherical 25
regenerate exhaust 144
spinel solid solution 161
to thermal deactivation, resistance of noble metal 24
C_2H_4 by base metal catalysts, oxidation of CO and 178
Chemistry, support 32
Chemical reactions in exhaust gas 43
Chemically pure aluminas 150
Chromatographic separation, gas .. 38
Chromite nickel aluminate, copper 164
Classes of spinels 175
Clean Air Act 1
Combustion products of light gases 22
Comparison of platinum and base metal oxidation catalysts 72
Composition on catalyst durability, effect of catalyst 27
Composition, optimum metals loading and 137
Carbon monoxide
and C_2H_4 by base metal catalysts, oxidation of 178
control 82
conversion by base metal catalyst, effect of catalyst thickness on 126
conversion for platinum kinetics, effect of catalyst layer thickness on 127
oxidation 187
activity vs. palladium 137
Contamination, degradation of NO_x catalysts by 85
Contaminants on monolithic NO_x catalysts, distribution of 95
Contaminant poisons, distribution of 97
Control catalysts, degradation in automotive emission control .. 103
Control catalysts, oxidative automotive emission 133
Control, CO 82
Conversion
by base metal catalyst, effect of catalyst thickness on CO .. 126

INDEX

Conversion *(continued)*	
CO	124
efficiencies, warm-up	78
for platinum kinetics, effect of catalyst layer thickness on CO	127
(TWC) catalyst, three-way	3
Converters, catalytic	74
Co_3O_4 supported on ZrO_2	189
Copper	51
aluminate	157
chromite–nickel aluminate	164
chromite supported on ZrO_2	187
ferrite	175
manganate	173
–nickel extrudate catalyst	88
Corundum, transformation of an aluminum hydroxide to	152

D

Deactivation	173
resistance of noble metal catalysts to thermal	24
Decomposition of transition metal sulfates	181
Degradation	
in automotive emission control catalysts	103
of NO_x catalysts by contamination	85
by poisons, catalyst	85
Deoxiding temperatures for metals in auto exhaust	51
Desulfiding temperature in auto exhaust	48
Detection of NO_x, thermocatalytic	14
Diffraction, x-ray	166
Dilution, effect of catalyst bed	141
Distribution	
between spheres, metals	140
of catalytic material on support layers in automotive catalysis	116
of contaminants on monolithic NO_x catalysts	95
within the catalyst sphere, metals	141
of contaminant poisons	97
Dual-bed method	2
Durability, effect of catalyst composition on catalyst	27
Dynamometer testing	90

E

Effectiveness factor	117
Efficiencies, warm-up conversion	78
Electron microprobe analysis	109
Electron probe studies	143
Emission(s)	
carbon monoxide	30
catalyst test unit, schematic of automotive	134

Emission(s) *(continued)*	
control	
catalysts, degradation in automotive	103
catalysts, oxidative automotive	133
platinum catalysts for exhaust	54
FTP	76
hot cycle	77
Engine aging of catalysts	3
Engine load, lead poisoning as function of	65
Environmental Protection Agency requirements	29, 144
Environmental Protection Agency test	33
EPM line scans	88
Equilibrium composition of auto exhaust	45
Ethane oxidation activities, carbon monoxide ethylene and	180
Ethylene, and ethane oxidation activities, carbon monoxide	180
Exhaust	
catalysts, regenerate	114
deoxiding temperatures for metals in auto	51
desulfiding temperature in auto emission control, platinum catalysts for	48
	54
equilibrium composition of auto gas	45
catalytic reduction of NO_x emissions in auto	244
chemical reactions in	43
composition of simulated	43
tests for three-component thermocatalytic analyzer	23
thermodynamic interaction between transition metals and auto	39
transition metals in auto	49
vehicle	23
External mass transfer	128

F

Federal Emission Standards, 1975	28
Fisher-Tropsch synthesis	47
Flame sources, flue gases from	23
Flue gases from flame sources	23
Fluorescence, x-ray	96
FTP emissions	76
Fuel	
additives as catalyst poisons, oil and	57
effect of lead content in	58
lead	114

G

Gas	
catalytic reduction of NO_x emissions in auto	244
chromatographic separation	38

Gas *(continued)*
  composition of simulated exhaust 34
Gases, combustion products of light 22

H

HC oxidation 81
HNO_3 and NO_2–N_2O_4, studies of
  gaseous 21
Honeycomb(s)
  activity tests on 187
  as a substrate 178
  parameters required 181
Hot cycle emissions 77
Hydrocarbon(s)
  activity 187
  oxidation
    activities of various oxides ... 179
    activity 138
    of paraffinic 25

I

Instability of aluminas, remedies for 155
Interaction between transition metals and auto, thermodynamic 39
Isothermal monolithic reactor 125

K

Kinetics, effect of catalyst layer thickness on CO conversion for platinum 127

L

Lead 51, 85, 109, 134
  content in fuel, effect of 58
  fuel 114
  and phosphorus additives,
    poisoning powers of 67
  and phosphorus compounds,
    catalyst poisoning by 54
  as poisons, phosphorus and 57
  poisoning, effect of catalyst
    temperature on 65
  poisoning as function of engine
    load and zinc 101
  and zinc 101
Lattice structure of transition
  aluminas 153
Line scans, EPM 88

M

Mass transfer, external 128
Mechanisms of poisoning 70
Metal
  catalysts, oxidation of CO and
    C_2H_4 by base 178
  catalysts to thermal deactivation,
    resistance of noble 24
  mobility 90

Metal *(continued)*
  oxidation catalysts, noble 24
  oxides, transition 161
  sulfates, decomposition of
    transition 181
  sulfides 49
  sulfiding, thermodynamics of .. 86
  vs. platinum catalysts, base ... 83
Metals
  and auto exhaust, thermodynamic
    interaction between transition 39
  distribution between spheres .. 140
  distribution within the catalyst
    sphere 141
  in auto exhaust, deoxiding
    temperatures for 51
  in auto exhaust, transition 49
  loading and composition,
    optimum 137
  pellets for use with base 178
  thermodynamics of noble 43
Metastability of transition aluminas 157
Microdistribution 95
Microprobe analysis, electron 109
Microscopic studies 165
Mobility, metal 91
Monolithic
  catalysts 32
  preparation of 25
NO_x catalysts, distribution of
  contaminants on 95
  oxidation catalysts, platinum ... 54
  reactor, isothermal 125
Monolith(s)
  and catalyst porosity 169
  vs. spheres 135
  washcoated 94

N

N-2 catalyst 7
NH_3 in air using vanadia–alumina
  catalyst, interaction of NO with 15
Nickel 51
  extrudate catalyst, copper 88
  platinum-promoted 32
NiSO 89
Nitric oxide 164, 171
  conversion 170
  disappearance 33
  reaction stoichiometry, ammonia 37
  removal activities 35
  response to oxides of nitrogen
    other than 21
Nitrogen oxides 32
Nitrogen other than nitric oxide,
  response to oxides of 21
NO_2–N_2O_4, studies of gaseous
  HNO_3 and 21
NO reduction, effect of catalyst
  aging on 159
NO with NH_3 in air using vanadia–
  alumina catalyst, interaction of 15

INDEX

Noble metal 51
  catalyst, base metal 96
  catalysts to thermal deactivation,
    resistance of 24
  oxidation catalysts 24
  thermodynamics of 43
NO_x
  catalysts 39, 51, 189
    automobile exhaust 85
    by contamination, degradation
      of 85
    distribution of contaminants on
      monolithic 95
    control processes 23
    emissions in auto exhaust gas,
      catalytic reduction of 244
    removal catalysts 32
    reduction catalysts 1

O

Oil and fuel additives as catalyst
  poisons 57
Optimum metals loading and
  composition 137
Oxidation
  activity 163
  carbon monoxide, ethylene,
    and ethane 180
  hydrocarbon 138
  of various oxides, hydrocarbon 179
  vs. palladium content, CO .. 137
  1-butene 19
  catalysts
    comparison of platinum and
      base metal 72
    noble metal 24
    platinum monolithic 54
    carbon monoxide 137, 187
    of CO and C_2H_4 by base metal
      catalysts 178
    HC 81
    of paraffinic hydrocarbons 25
    with spinel solid solid solutions,
      propylene 169
Oxidative automotive emission
  control catalysts 133
Oxides
  hydrocarbon oxidation activities
    of various 179
  nitrogen 32
  of nitrogen other than nitric
    oxide, response to 21
  rare earth 158
  reaction stoichiometry,
    ammonia–nitric 37
  removal activities, nitric 35
  transition metal 161
  with carriers, reactions of
    activated 159

P

Palladium 137, 143
  catalysts 106, 114

Palladium (continued)
  content, CO oxidation activity vs. 137
  mixture, thermally-optimized
    platinum– 28
  and platinum resistance to sulfur
    and thermal aging 139
  sintering of platinum and 107
Paraffinic hydrocarbons, oxidation
  of 25
Parameters required, honeycomb .. 181
$PbSO_4$ 88
Pellets for use with base metals .. 178
Performance, warm-up 76
Phase transformations 158
Phosphorus 85, 101
  additives, relative poisoning
    powers and 67
  compounds, catalyst poisoning by
    lead and 54
  and lead as poisons 57
Photochemical smog 1
Platinum 137, 143
  and base metal oxidation cata-
    lysts, comparison of 72
  catalysts 114
  base metal vs. 83
  for exhaust emission control .. 54
  and palladium, sintering of 107
  poisoning of 62
  kinetics, effect of catalyst layer
    thickness on CO conversion
    for 127
  monolithic oxidation catalysts .. 54
  –palladium mixture, thermally
    optimized 28
  -promoted nickel 32
  resistance to sulfur and thermal
    aging, palladium and 139
Poisoned catalyst, Arrhenius plot
  for a 111
Poisoning
  catalyst 107
  effect of catalyst temperature
    on lead 65
  as function of engine load, lead 65
  by lead and phosphorus com-
    pounds, catalyst 54
  mechanisms of 70
  of platinum catalysts 62
  powers of lead and phosphorus
    additives, relative 67
  studies 173
  sulfur 86
Poisons
  catalyst degradation by 85
  deposition on the catalyst 68
  distribution of contaminant ... 97
  oil and fuel additives as catalyst 57
  phosphorus and lead as 57
Porosity, monolith and catalyst ... 169
Preparation of monolithic catalysts 25
Preparation of spherical catalysts .. 25
Probe studies, electron 143

Properties, catalyst 74
Properties of carriers 149
Propylene 48, 171
  oxidation with spinel solid
    solutions 169
Protection Agency (EPA), require-
  ments, Environmental 144

R

Rare earth oxides 158
Reactions of activated oxides with
  carriers 159
Reactions in exhaust gas, chemical 43
Reactor, isothermal monolithic ... 125
  activity 164
  catalyst, NO_x 1
  effect of catalyst agin on NO .. 159
Regenerate exhaust catalysts 114
Removal activities, nitric oxide .. 35
Removal catalysts, NO_x 32
Required, honeycomb parameters 181
Ruthenium catalyst, supported ...   2

S

Scans, EPM line 88
Selectivity, ammonia 32
Selectivity, catalyst 32
Sensor carburetor system 3
Separation, gas chromatographic .. 38
Silica 157
Simulated exhaust gas, composition
  of 34
Sintering 152, 179
  of platinum and palladium 107
  of ZrO_2 washcoat 184
Smog, photochemical 1
Solution catalysts, spinel solid ... 161
Solid solution catalysts, spinel ... 161
Solid solutions, propylene oxidation
  with spinel 169
Space velocity 170
Sphere, metals distribution within
  the catalyst 141
Spheres, metals distribution
  between 140
Spheres, monoliths vs. 135
Spherical catalysts, preparation of 25
Spinel
  compounds 165
  solid solutions, propylene
    oxidation with 169
  solid solution catalysts 161
Spinels, classes of 175
Stable catalyst carriers, thermally 147
Stability, thermal 105
Stability of various aluminas 153
Standards, 1975 Federal Emission 28
Structure of transition aluminas,
  lattice 153
Substrate, honeycomb as a 178
Sulfates, decomposition of transition
  metal 181
Sulfiding, thermodynamics of metal 86
Support
  chemistry 32
  layers in automotive catalysis,
    distribution of catalytic, ma-
    terial on 116
  on acidity, effects of the 35
Support, properties of catalyst ... 33
Supported ruthenium catalyst   2
Sulfides, metal 49
Sulfur 85, 109, 134, 188
  dioxide 22, 173
  poisoning 86
  and thermal aging, palladium
    and platinum resistance to 139
Surface areas 183
Synthesis, Fisher-Tropsch 47
Synthetic means of aging catalysts 145

T

Temperature
  in auto exhaust, desulfiding ... 48
  catalyst inlet   8
  on lead poisoning, effect of
    catalyst 65
  for metals in auto exhaust,
    deoxiding 51
Test unit, schematic of automotive
  emissions catalyst 134
Testing, dynamometer 90
Tests on honeycombs, activity ... 187
Thermal deactivation, resistance of
  noble metal catalysts to 24
Thermally-optimized, platinum–
  palladium mixture 28
Thermocatalytic analyzer, vehicle
  exhaust tests for three-compo-
  nent 23
Thermocatalytic detection of NO_x 14
Thermodynamic interaction be-
  tween transition metals and
  auto exhaust 39
Thermal aging, palladium and plat-
  inum resistance to sulfur and 139
Thermally stable catalyst carriers 147
Thermal stability 105
Thermodynamics of metal sulfiding 86
Thermodynamics of noble metals .. 43
Thiele analysis 116
Thickness on CO conversion by
  base metal catalyst, effect of
  catalyst 126
Thickness on CO conversion for
  platinum kinetics, effect of cat-
  alyst layer 127
Three-way conversion (TWC)
  catalyst   3
Transfer, external mass 128
Transformation of an aluminum
  hydroxide to corundum 152
Transformations, phase 158
Transition
  aluminas, metastability of 157

INDEX

Transition *(continued)*
  metal
    and auto exhaust, thermodynamic interaction between ... 39
    in auto exhaust 49
    oxides 161
    sulfates, decomposition of .. 181
  Tropsch synthesis, Fisher– 47

V

Vanadia–alumina catalyst, interaction of NO with NH_3 in air using 15
Vehicle exhaust 23
Vehicle exhaust tests for three-component thermocatalytic analyzer 23
Velocity, space 170
Volume on catalyst deterioration, change in catalyst 108

W

Warm-up
  characteristic of catalyst 8
  conversion efficiencies 78
  performance 76
Washcoated monoliths 94
Washcoat, ZrO_2 as 187

X

X-ray
  diffraction 166
  analysis 88
  fluorescence 96

Z

ZrO_2
  Co_3O_4 catalyst supported on ... 189
  copper chromite supported on .. 187
  washcoat 187
  sintering of 184
Zinc 85
  lead and 101

The text of this book is set in 10 point Caledonia with two points of leading. The chapter numerals are set in 30 point Garamond; the chapter titles are set in 18 point Garamond Bold.

The book is printed offset on Danforth 550 Machine Blue White text, 50-pound. The cover is Joanna Book Binding blue linen.

Jacket design by Sharri Harris.
Editing and production by Spencer Lockson.

The book was composed by the Mills-Frizell-Evans Co., Baltimore, Md., printed and bound by The Maple Press Co., York, Pa.